Narrative and Technology Ethics

"Narrative & Technology Ethics represents a major contribution to the contemporary philosophical debate on technology. It will also represent a formidable instrument for practitioners, and stakeholders in general. The authors have indeed the great merit of translating the narrative ethics of technical practice into an elegant and easy-to-apply method for practicing ethics in research and innovation."

—Alberto Romele, *Associate Professor of Philosophy of Technology at the ETHICS Lab of Lille Catholic University, France*

Wessel Reijers • Mark Coeckelbergh

Narrative and Technology Ethics

palgrave
macmillan

Wessel Reijers
European University Institute
Florence, Italy

Mark Coeckelbergh
Philosophy institute
Universität Wien
Wien, Wien, Austria

ISBN 978-3-030-60274-1 ISBN 978-3-030-60272-7 (eBook)
https://doi.org/10.1007/978-3-030-60272-7

This Palgrave Macmillan imprint is published by the registered company Springer Nature Switzerland AG.
The registered company address is: Gewerbestrasse 11, 6330 Cham, Switzerland

To Irene, Hubert, and Valentine
To Wim

Acknowledgements

This book is a testimony to the fact that academic research is not a one man show, but rather a collective effort—even an "acting in concert", as Hannah Arendt might have put it. It is embedded in a philosophical and practical discourse that stretches back for millennia, and it has been shaped through constant interlocution between colleagues, friends, and family. We have had a great time writing this book and working constructively with the editors of Palgrave Macmillan; in particular Lauriane Piette who has been very helpful in guiding the journey from book proposal to publication.

The book is partly built on previous research that has been published in different journals. Chapter 3 contains a section that is a reworked version of Reijers, W. (2019). "Beyond Postphenomenolgy: Ihde's Heidegger and the Problem of Authenticity". *Human Studies* 42, 601–619. Chapter 4 contains reworked sections of two papers: Coeckelbergh, M., & Reijers, W. (2016). "Narrative Technologies: a philosophical investigation of the narrative capacities of technologies". *Human Studies*, *39*(3), 325–346; and, Reijers, W., & Coeckelbergh, M. (2018). "The Blockchain as a Narrative Technology : Investigating the Social Ontology and Normative Configurations of Cryptocurrencies". *Philosophy & Technology*, *31*(1), 103–130. We are thankful that the writing of this book was supported by the European Research Council (ERC) under the European Union's

Horizon 2020 Research and Innovation Programme [Grant Agreement No. 716350].

In the course of writing the materials for this book, we have gained valuable input from colleagues, most notably in the research group "Ricoeur and Technology". We would particularly like to thank Alberto Romele, Bert Gordijn, Declan O'Sullivan, Fiachra O'Brolchain, Quinn Dupont, David Lewis, Renaat Verbruggen, Marinus Ossewaarde, Vincent Blok, Morshed Mannan, Primavera de Filippi, Johnny Søraker, Jules Lepoutre, Federico Tomasello, and Liav Orgad. Beyond our professional life, our thinking has also been nurtured by philosophical conversations with good friends. A special thanks goes to Hilde de Leeuw, Liesbeth Stam, Beer Sijpesteijn, and Paul Steen, for sharing a passion in philosophical but also very "earthly" thinking.

Finally, we could not have found the perseverance to write and especially finish the book without the continuous encouragement and help from those closest to us. For Wessel: Lisa, for providing a loving home, a spirited connection of souls, and a warm heart for both me and Arthur, our little one; and also, Hubert and Valentine, for being the anchors in life. I am grateful for having Mark as a co-author of this book, who has been a good friend and mentor. For Mark: Sabine and my wonderful children Lotte and Arno; my parents and my brother Wim; my friends, for joining me on my path. Warm thanks to Wessel, a great person and excellent lead author of this book.

Contents

List of Tables

1

Introduction

Narrative and Technology Ethics appears on the scene in a time of great disruption. Across the globe, societies are facing the crisis of the Covid19 pandemic, as well as its looming aftermath. The pandemic is not only a crisis of economics and politics, but also of technology. Indeed, technologies feature heavily in accounts of the virus' spread and our responses to it. On the downside, technology has been a major factor in the rapid dispersion of the virus; through mobility networks that have been enabled by vast infrastructures and machines. On the upside, technology has enabled rapid response. Despite the sometimes-grave mistakes that have been made by countries in tackling the pandemic, there can be little doubt that without the technologies to quickly disseminate electronic messages, trace populations, and heal patients, death tolls would have been immensely higher. Compared to earlier global pandemics, such as the black death and the Spanish flue, the world has done a remarkable job in turning the tide. Technology has played a central role in the Covid19 stories and is likely to keep playing this role in the virus' aftermath; as apps are being built to track people who get sick, new medical innovations are in the making, and also the understandable fear for technology might increase.

W. Reijers, M. Coeckelbergh, *Narrative and Technology Ethics*,
https://doi.org/10.1007/978-3-030-60272-7_1

Thus, Covid19 shows how technologies play an important role in the stories that shape our understanding of where we are and where we are heading. Yet, stories concerning technology have been around for millennia. For instance, considering how much it appears in their myths; the ancient Greeks might well have loved technology. Yet, at the same time, they were utterly pessimistic about its consequences for human lives, or at the very least warned about the limits of what technology can do. In the Egyptian myth about the invention of writing, as told by Plato, the technology of writing is said to produce forgetfulness in the minds of readers. Prometheus gets punished when he brings the technology of fire to the mortals. Icarus gets artificial wings but the wax that holds his feathers together melts when he flies too high and he dies. And Pandora is also high tech: 'made, not born', she is an artificial woman or automaton, ordered by Zeus and made by Hephaestus. Disguising evil as beauty, her mission is to release suffering, death and disease upon humankind. Pandora's myth reveals the normativity of technology in its historical context, for it is highly charged with the negative view of femininity in a patriarchical society.

Quite similar to the Greeks, we now ask: *How to live well with technology?* This question seems to be at the top of everyone's mind. Whether one opens the newspaper, turns on the TV, or scrolls through Twitter, one is constantly confronted with the most recent breakthroughs in AI, the launch of yet another app, and the perils of the Chinese Social Credit System. The world of today is literally littered with *stories* about technology, about the manifold ways in which we think and express our thoughts about the giant artificial world we—together—have created. These stories range from those of today's optimists, who rage about the possibilities for constructing the stone of (eternal) life that Medieval alchemists dreamt of; uploading one's mind to the machine and living forever; to the stories of today's pessimists, who see the same world as one that brings doom to mankind, through never-ending pollution and poisoning of the soul. Just a single peek at the news confirms this tendency. Smart technologies make our homes safer and more comfortable; cryptocurrencies hold the promise of "banking the poor"; and new medical technologies might cure most deadly diseases in the foreseeable future. At the same time, home assistant "Alexa" represents the relentless penetration of the household by giant tech companies; social media are fuelling worldwide spread of political hatred; and global industrial activities threaten to turn the earth into an inhabitable place,—threatening our very common home. As Bernard Stiegler (1998) rightfully observed, technology is similar to what the Ancient Greeks called a *pharmakon*: it is both the poison and the cure (or perhaps rather: neither the poison, nor the cure). At current, we are engaged in a global conversation, a web of stories, to answer the central question of how to live well with such a phenomenon.

But who are the protagonists and antagonists of these stories? Is this just about humans, for example about humans using technology or humans defending themselves against the bad effects of technology? If there is one thing that philosophers of technology have shown us, it is that it is not simply "us" against the "machines". We cannot simply abandon technology and retreat to a pre-technological world; both because this "world" has never existed and because technology constitutes a condition of our existence. We are tool-making beings, in the words of Georg Simmel (1900). Exactly in this realisation, we find the key to seeing that we do not only do things with technologies, but *together* with them. Technologies are an essential part of the stories we tell, and they have a certain kind of agency; whether as "actants", or as "mediating" artefacts. But why, one might ask, is it of any importance that technologies have agency in answering the question of how to live with them? Can we not simply use technologies in a better way? As long as we use the hammer to hammer, and not to hit each-other on the head, we should certainly be fine? However, the problem is that this description of human-technology relations does not quite fit the phenomena. In our experience and in what we do, technologies co-act our everyday choices and practices. In terms of a mantra that never seems to get old: "technology is not neutral" (but: what is it, instead?). Technologies persuade us, teach us, invite us, inhibit us, harm us, and thereby passively or actively contribute to the ethical choices we make and actions we engage in. Understanding this dynamic will help us make, use and govern technologies in ways that are conducive of the good life. Without understanding this dynamic, we seem to be lost.

This book departs from the idea that this dynamic is captured by the notions of *technical practice* and *narrative*. We start with the idea of technology-in-practice, or rather technical practice. If philosophy of technology has an object of study, it is *not* merely the actual thing, or artefact, but rather the thing-in-action. On this point, we agree with the latter-day giants of philosophy of technology like Don Ihde (1979), who argue that *praxis* should be the focal point of our study. For this reason, philosophy of technology should be "empirically informed": it should study everyday practices to understand the role that technologies play in society. However, as we will see, we believe that the questions "what is a technical practice" and "how can we know about it", have so far not been answered in a satisfactory way. This is because philosophers have not paid sufficient attention to the things that are most obvious at the surface level: the stories about new technologies told in newspapers and blogs, the testimonies of users on social media, the sharing of coding experiences at a "hackathon", the tech-policies written by lawyers and politicians. These all reveal something Alisdair MacIntyre (2007) has said years ago but philosophy of technology is only engaging with now: that practices are made intelligible through stories, through *narrative*.

It is in the narrative mode that we explain our actions-with-technologies; that their "agency" is revealed. It is in the narrative mode that technical practices gain significance; that riding a bicycle becomes a social or political statement; that operating a military drone in a warzone becomes conducive or obstructive of certain virtues, of courage or friendship; and that smart architectures afford us a sense of freedom or domination. And, it is in narratives that we find the clearest "reflection" of technical practices; the *topos* where we can read how technologies mediate our actions. The major idea of his book is that narrative holds the key to our understanding of technical practices; both in explicating how narrative structures constitute a horizon of our understanding of the role of technical things in practices, as well as in showing how technical things mediate the narrative structures that govern our understanding of practices. In other words, this book depicts a dialectic between our *understanding* of technical practices and the *explanation* of these practices through narrative structures.

In science, technology, and society studies (STS), it is nothing new to say that people make sense of the use of technologies in their everyday lives by telling stories. For instance, Trevor Pinch and Wiebe Bijker (1984) have explored how social groups use different interpretations of a technology to negotiate about its design and use; a process which moves towards interpretative "closure" that stabilises a technology. When a new medical technology like a prosthesis is developed, patient advocacy groups negotiate with medical professionals, hospitals, and government actors negotiate about its possible uses. Similarly, our visions of new technologies are highly influenced by stories in science fiction; from Asimov's robot laws, to the cyber-Utopia of Star Trek, to the Black Mirror TV series. Erroneous as many of these fictional accounts may be, not to mention the transhumanist fiction of the "singularity" (Kurzweil 1999), which has had a strong impact of the ideology of innovation in Silicon Valley; they do "prefigure" people's understanding of technical practices and the extent to which they persuade, seduce or frighten us. As Sheila Jasanoff (2015) explains, "sociotechnical imaginaries"—collectively imagined forms of social life that are incorporated in the design of scientific and technological projects—drive technological development and implementation. Furthermore, in philosophy of technology, David Kaplan (2006) has argued that we use narrative to make sense of technologies and has suggested that the work of Paul Ricoeur offers "a narrative theory of interpretation for making sense of all the different ways that technologies figure into our lives" (Kaplan 2006, p. 50). He offers the example of stories of technological development, which help us to understand the design of a technology and reveals hidden actors and motives, and stories of identity-formation, which explain how technologies figure in our lives.

We want to step beyond these initial intuitions and investigate how technical practices are not only "prefigured" by the manifold of stories that are told about them (that is, play a passive role in our narratives about them), but also actively "configure" and "refigure" these stories through our interactions *with* them. It is at this point that STS meets philosophy of technology, and two distinct paradigms are revealed. On the one hand, contemporary philosophy of technology has occupied itself predominantly with the "materiality" of things and their role in technical practices. For instance, it has considered how the materiality of for instance a pair of virtual reality glasses mediates our experience and makes the technology appear as "embodied" (Kudina and Verbeek 2018). Material things "do" something to us by mediating our experiences and our actions. Although paying more attention to language and narrative, Bruno Latour (1994) similarly offers an account of material "scripts" that are not constituted by words arranged in a text but by material things that operate as semiotic elements; instantiating material "programmes" that inhibit and invite us. On the other hand, STS has, partly due to its greater affinity with analyses of texts and discourses in history, sociology, and media studies, focused on the role of human discourse in the shaping of technical practices. It has thereby paid insufficient attention to "the things themselves".

Going *beyond* both these strands of theory and drawing on narrative theory, this book puts a new and far-reaching thesis forward: that technologies, similar to texts, novels, and movies, "tell stories" by configuring characters and events in meaningful syntheses. As with most philosophical theses, this is one that reflects the current of its time. Why does it seem that technologies increasingly do not only do things *to* us, but *with* us, notably in the form of discourse? In line with Andrew Pickering's striking observation (2010, p. 7), we have surpassed the age of technologies such as the bridge—that "just stand there"—and entered the age of the "adaptive machines", with a "disturbing life-like quality". Around us are myriad systems and devices that watch us, hear us, and engage with us; from smartphones to self-driving cars. Technologies are increasingly engaging with us in active ways, sensing our behaviours, our emotions, and even our thoughts, and responding to them through processes of internal configuration. Moreover, they do so in ways that are often very textual and expressive; using natural language and other modes of communication conducive to human interpretation. At the same time, they remain at a certain distance from us and retain or even expand aspects of technical practices that are impenetrable; where the stories they co-act with us do not "add up" or leave us with a sense of estrangement.

Narrative theory enables us to understand both the structures of these mediations and their limitations. On the one hand, it helps us understand how humans and technologies co-act technical practices insofar as they remain conducive to "human time", open to our understanding. As such, it can reveal how technologies "emplot" different characters and events in a meaningful whole. On the other hand, narrative theory helps us to explore the limits of technical practices, to see where technologies resist interpretation and become increasingly "autonomous". It can do so by analogy of limits of meaning in a text: the ways in which a textual practice (which, essentially, is a technical practice) can lose its human sense of time and distanciate itself progressively from the world of action.

In our endeavour to use narrative theory in constructing an account of technological mediation, we draw extensively from the work of Paul Ricoeur. Ricoeur has been one of the most influential philosophers in the last century working on critical hermeneutics, and his oeuvre spans a wide range of themes; from linguistic analysis, to ethics, to political theory. He also developed one of the most comprehensive and philosophically informed narrative theories, which is the main reason why we will take up his work in developing our thesis concerning the active narrating role of technology. We will also use Ricoeur to think about the relation between technology and society. We thus take up Kaplan's recommendation that Ricoeur's work has much to offer to philosophers of technology but move beyond this by showing how the link can be made. In addition, we take inspiration from, and engage with, another body of theory that has its roots in Aristotelian philosophy: virtue ethics. This is not only an increasingly popular normative theory in contemporary philosophy of technology but offers much in combination with a narrative theory of technology in the form we propose. We simultaneously position our arguments within philosophy of technology, engaging with mediation theory as developed by Ihde and Verbeek, who have successfully initiated an influential but not unproblematic way to conceptualize the non-instrumental role of technology. Finally, we engage in an ongoing dialogue throughout this book with the new field of "digital hermeneutics", which investigates how human interpretation extends beyond the mind into the realm of digital machines; which have their own interpretative capacities (Romele 2019; Romele and Terrone 2018).

Taken together, the book develops a philosophical theory of technology, thereby containing a two-fold contribution. First, it develops a theory of technology and corresponding ethics of technology that draws from narrative theory. Second, the book positions Ricoeur's philosophical hermeneutics as a central point of departure for thinking about the role of technology in society. It thus aims to contribute to a central problematic in philosophy of technology but also offers some insights that may be useful to STS and responsible innovation theory, among other related fields.

Mediation, Virtue, Responsibility

In terms of the rationale for this book and the discussions we connect to, we intend to integrate three interconnected intentions that derive from different fields of study: philosophy of technology, ethics of technology, and responsible innovation. As its title, *Narrative and Technology Ethics*, conveys; we do not merely mean to offer a novel theory of technology but also aim to develop a framework to reflect on the ethics of technical practices. We believe that, in a sense, every theory of technology is already an ethical theory in disguise, insofar as it deals with the way technologies mediate our normative beliefs, attitudes, and behaviours. However, most theories of technology have been developed separately from the ongoing discourse on normative ethics, which is something we want to remedy in this book. We intend to make narrative relevant descriptively, normatively, and also prescriptively, by connecting with the developing body of literature that aims at shaping the practise of ethics in research and innovation; the *site* where technologies are conceived of and from where they are brought into the world. As such, we intend this book not only to be relevant for philosophers and scholars from other fields studying technology such as STS, but also for practitioners in research and innovation (engineers, computer scientists, etc.), as well as for policy makers.

Let us further articulate three guiding intentions, which also helps us to further unpack the problems, discourses, and thinkers we engage with. The *first intention* of this book is to gain a better understanding of the phenomenon of "technological mediation", which has arguably been the focal point of philosophy of technology in the last decades. How to understand this phenomenon? At the surface level, people already grasp the meaning of this concept by discussing the role of technologies in society. For instance, people engage in lively online discussions about the way in which augmented reality glasses change their wearer; whether they might turn someone from a law-abiding citizen into a "glasshole" (Kudina and Verbeek 2018). Similarly, journalists and commentators discuss how the introduction of iPads in school classes produce different teachers and students (March 2014), or how new advances and biomedical technologies create new moral choices, such as the choice to intentionally have a baby with a disability (Teather 2002). What these examples show, is that humans already have a pregiven "grasp" of the idea of technological mediation. It is true that, as philosophers of technology often lament, many people do not pay attention to the "non-neutrality" of technology, but this seems more a question of focus and awareness than of general receptiveness. In other words, people are perfectly capable of grasping the idea of technological mediation, they just do not always pay attention to it (or worse, actively neglect it).

What then, one might ask, is the role of philosophers, apart from "raising awareness"? We think this role is exemplified by philosophers such as Martin Heidegger and Hannah Arendt. Perhaps more than any other philosophers of technology, they have argued from different positions that the question of technology is a question concerning a *mode of being*. We understand that technical practices mediate our experiences and actions, but the question is *how* they do so. Technologies-in-action enable augmented or new modes of being that have particular structures. A beautiful and at the same time terrifying illustration of this is the manufacture of pieces of so-called "wampum", which are colourful beads made out of shells, by European settlers in Northern America, during the seventeenth and eighteenth centuries (Graeber 1996). These objects were manufactured and used as money to trade with local populations. However, these local populations had never regarded these objects—which they had used for hundreds of years—as money and had to adopt their lives to this new reality of money making and trading; with often terrible consequences for their communities. Quite suddenly, they had to engage in economic calculations and market thinking concerning aspects of their life that no market had ever touched. This example shows that a technology such as money, which Georg Simmel designated as the "perfect tool" (1900), mediates technical practices that bear the capacity to constitute a new mode of being; a *technical* mode of being. The task of philosophers is to shed light on this phenomenon.

So again, we have to ask: what is technological mediation? Philosophers define it in a variety of ways, as technology *influencing* (Dorrestijn 2011), *co-shaping* (Verbeek 2005), or *translating, composing*, and *delegating* (Latour 1994) human experience and action. Borrowing from narrative theory, we will instead characterise it as the *configuration* of practices by technologies. As we will discuss in more detail in Chap. 2, this notion of configuration is analogous to the way in which a narrative that we engage with in a text configures our understanding of the world. That is, a text has a particular structure in which events and characters are organised, which generates a sense of time and puts what is narrated at a certain distance from the "world of action" of the characters. For instance, through reading Kafka's novels, a reader's understanding of the social world and the role of bureaucracy is *configured*, *refiguring* everyday actions into a new mode of being with a distinct temporal and spatial structure. The text has changed the reader's understanding of simple actions such as checking in and out of her office, as well as filing requests for information with the authorities; which appear to her with an increasing sense of impersonality and distance. Similarly, using a technology such as the Strava app that tracks a user's running progress, route, and patterns, transforms the technical practice of "running" and reveals a new sense of temporality and spatiality to the user.

In line with Ricoeur, we maintain that a text can be considered as a model of meaningful action, and by extension, as a model for meaningful technical practice (and consequently, as a source for understanding the *limits* of the meaningfulness of technical practices). A text is a useful object of analysis because its structures are more or less transparent; it reveals structural elements (e.g., grammar, sequencing) that mediate the temporal and spatial aspects of the reading experience. It also shows how these aspects can be transformed to make a text increasingly "inhuman", blocking access to interpretation and therefore to understanding. Treating technology in an analogous way, we are able to outline the character of the mediating structures of a technical practice; both being conducive to human understanding and resisting it. Not the least, this is due to the fact that technical practices in empirical research of technologies are captured in textual form; *as stories*. Whether it concerns hotel room visitors using door keys (Latour and Porter 1993), ultrasound and its relation to parents and the unborn child (Verbeek 2008), or scientists using brain imaging technologies (De Boer et al. 2018); empirically informed philosophy of technology captures the technical practices it is interested in as stories. Yet, we have to distinguish between story and narrative structure: a story, as we understand it, is a narrative structure captured in a linguistic form (a text or utterance), whereas narrative more broadly understood captures any phenomenon that is conducive to narration. As such, the technical practice of driving a car might contain (or rather, "have") a narrative structure but *is* not strictly speaking a story.

The *second intention* of this book is to link our account of technological mediation with an ethics of technology, in particular to avoid divorcing our discussions on the normativity of technical practice from the ongoing discourse in normative ethics. The need for a move from technical mediation to its ethical impacts is motivated by growing and widespread concerns about the way in which new technologies influence human morality. On the individual level, new technologies such as home assistants challenge our autonomy, persuading us to do certain things (e.g., to buy a new item on Amazon) or even think in certain ways (Frischmann and Selinger 2018, p. 3). On the inter-personal level, platforms that focus on relationship-building and dating heavily impact virtues of empathy and friendship (Frischmann and Selinger 2018, p. 158). On the institutional level, global public discourse is increasingly challenged by new structures of broadcasting and moderation; giving rise to the spread of misinformation (Frischmann and Selinger 2018, p. 279). These developments and many others indicate that the phenomenon of technological mediation prompts inquiries into the "moral" mediation of technologies (cf., Verbeek 2011). Hence, it is not enough to merely have an approach to understand the structure of technical practices, but this needs to be complemented by a corresponding way to evaluate the normative impacts of these practices. In setting up this approach, we will align our thinking with the newly emerging discourse on "virtue ethics of technology", while taking our primary cues from MacIntyre and Ricoeur, focusing on the latter's "Little Ethics".

Our choice of taking virtue ethics as the basis for our ethical approach is not coincidental, given that we approach the question concerning technology as being linked to technical practice. By developing our concept of technical practice from the basis of Ricoeur's critical hermeneutics, we incorporate a notion of practice that is closely related to the idea of practice in MacIntyre's (2007) virtue ethics. Like Ricoeur's narrative theory, the latter is also inspired by Aristotle and is also informed by the intuition that human beings have a shared existence. Inspired by both Ricoeur and

MacIntyre, we define a practice as a shared human activity that has an "internal" end, which is established through standards of excellence, which in turn are made intelligible in the *narrative* mode. When we consider the mediating role of technology, we can immediately observe that we will be concerned with the way in which technologies mediate these standards of excellence, and therefore are conducive of our block those human dispositions that are called the virtues. For example, if we consider Latour's classical illustration of the gunman (Latour 1994), the technology of a gun mediates the standards of excellence of certain practices (e.g., proportionality in defence for police officers); which are made intelligible through narrative structures (e.g., disclosed through new professional codes). In other words, it is inherent in our theory of technology that the normative concern requiring our attention will be the relation between technical practices and the virtues.

Moreover, we argue that virtue ethics is the most promising starting point for bridging the gap that currently exists between philosophical theories of technology and theories in normative ethics. The major contemporary approaches in philosophy of technology, Actor Network Theory, Postphenomenology, and Critical Theory of Technology, all at some point have come to clearly aspire to position themselves as *ethical* theories of technology. Latour, who initially positioned himself more as a descriptive social scientist, has turned to concerns of morality at a later point (cf. Latour and Venn 2002); Verbeek has developed an ethics of technology in *Moralizing Technology* (Verbeek 2011), and Feenberg has throughout his work been concerned with normativity; both in terms of politics and ethics (Feenberg 1999). Yet, none of these approaches has directly engaged with the ongoing discourse in normative ethics. Being concerned with technological mediation, these theories focus on the way in which technologies mediate the affordances, attitudes, or (cultural) habits that people have. As such, they are not primarily concerned with evaluating merely the intentions of users and designers—as a deontological approach would presuppose—or the consequences of technical practices—as a consequentialist approach would presuppose—but rather with the *way* in which technologically mediated action comes about. This, we argue, is most pertinently captured by a virtue ethics approach; because such an approach is concerned with the way in which practices are related to the moral character of an agent, and the dispositions that make up this character. In other words, a virtue ethics approach shows a good fit with the main theories of technology, as Shannon Vallor already remarked (Vallor 2016, p. 31).

The "revival" of virtue ethics—understood as a renewed interest in Aristotelian ethics in particular—has produced a great amount of scholarly works in the past decades. However, virtue ethics has only recently been introduced in the writings on ethics of technology. Charles Ess (2009) is one of the first scholars who sketched the outlines of a virtue ethics of technology and others have used virtue ethics in their ethical reflections on technology (cf. Bolsin et al. 2005; Coeckelbergh 2012). After these first explorations, Vallor's recent book *Technology and the Virtues* (2016) can be regarded as the first systematic and comprehensive account of a virtue ethics of technology approach. Virtue ethics has been a welcome addition to ethics of technology, because it offers an alternative, fruitful response to tensions in utilitarian and duty-based approaches that are very dominant in the field. Additionally, as Vallor argues, virtue ethics does not offer straightforward solutions to complex ethical issues as many approaches try to do, but instead provides a strategy for "cultivating the type of moral character that can aid us in coping, and even flourishing" (Vallor 2016, p. 10) with and under the challenging conditions of contemporary emerging technologies, such as surveillance technologies and human enhancement technologies.

We should note, however, that even though virtue ethics is our point of departure we do not take it as the only basis for an ethics of technology. In addition, we believe that a deontological viewpoint is needed, for a particular reason. That is, action—and in particular action-with-technology—always carries with it the capacity to exercise power over people and thereby the capacity to inflict (unintended) violence. Even when we design our technical practices in such a way that they are conducive of the good life, they might still exert (unintended) harm on certain people. For this reason, technical practices need to be limited in certain respects; they need to be subjected to appropriate laws. Even though Verbeek is right in claiming that we should emphasise "*how* to shape the interrelatedness between humans and technologies" (Verbeek 2011, p. 164—emphasis added), we should not forget to also draw boundaries; not between humans and technologies but *around* technical practices. We will therefore, in line with Ricoeur, add a moment of obligation to our virtue ethics of technology that aims at the good, in the form of a "sieve of the norm" (Ricoeur 1992, p. 170). This means that we will conceptualise how technical practices should be subjected to norms (laws, codes, etc.) aimed at eliminating their capacity to do violence.

The *third intention* of this book was to develop a method for practising ethics in innovation contexts, thereby drawing our theoretical discussions into the realm of practice. By doing so, we arrive at the *site* where technical practices are inaugurated into the world. This last turn is important because a general ethics of technology is still far removed from the concrete reality of research and innovation; of university labs in which new advances in robotics are worked on, of product teams in large tech firms in which new apps are developed that will affect the lives of millions of people, and of government owned research facilities that develop the latest weaponry to be deployed in conflict areas around the globe. Responsible innovation, understood as the promotion and implementation of ethical technical practices in innovation contexts, cannot be simply an application of normative theory, but also needs to be responsive to the innovation process. Innovation is classically conceptualised as a set of practices (1) based on conceptual (scientific) knowledge that (2) translates a certain conceptual idea into a technological design that is tested, prototyped and translated into an actual application, which (3) can subsequently be put into production and introduced to society at large (see e.g., Hauser et al. 2006; Kajikawa et al. 2008; Roberts 2007). This notion of innovation mostly corresponds to the "linear model of innovation" (Godin 2014), which is dominant but far from the only model of innovation. In relation to ethics, it is argued that most conceptualisations of innovation are insufficient because of their narrow focus on technological, economically motivated innovation rather than on innovation for the common good (cf. Blok and Lemmens 2015). To overcome this narrow focus, it is suggested that the concept of innovation should include non-technological novelties and alternative strategies of innovation that offer radically different types of stakeholder relations and responsibilities, such as peer-to-peer production. To accommodate these concerns, we characterise innovation in a broad way, as a process consisting of a set of practices aimed at (1) the invention and design of new material-cultural phenomena through (2) the use of techno-scientific knowledge and (3) the introduction and diffusion of these phenomena in society.

Our aim here is to complement and improve upon the extensive work done in developing methods for practising ethics in research and innovation (Reijers et al. 2017). This work has happened in parallel with a global response to the ethical concerns raised by new technologies; by activist groups, civil society organisations, and governments. First, these concerns have led to initiatives for responsible foresight: different types of impact assessment that aim at anticipating negative impacts of technical practices before they substantiate. For instance, organisations use tools like "Ethical OS" and the "Ethics Canvas" to anticipate typical risks that arise with the development of emerging technologies. Second, concerns have led to global initiatives for ethical design that take a particular ethical aim as their point of departure. Examples are the "Signal" app, which is designed to foster an idea of the good life in which privacy is protected, or "Democracy Earth", which is a platform designed to foster democratic engagement and decision-making. Third, global worries have led to new policies and regulations in corporate and governmental contexts that aim to limit the negative impacts of existing technical practices. Major examples in this respect are the both heavily criticised and applauded European General Data Protection Regulation (GDRP), but also controversial "soft" guidelines such as the ones proposed by the High-Level Expert Group of Artificial Intelligence at the European Commission.

At the academic front, broadly speaking three different kinds of methods have been developed that focus on practising ethics in innovation contexts. First, methods that focus on the anticipation of potential negative impacts of technological innovations on society. These methods propose foresight tools such as horizon scanning (cf. Brey 2012) or scenario building (cf. Boenink et al. 2010) to analyse technologies that are still early in the innovation process, and argue for probable future pathways of these technologies and the potential normative concerns that these pathways raise. Second, methods that aim to integrate ethical reflection in the design phase (e.g., prototyping) of a technological innovation, for instance by formulating stakeholder values that should guide the design process as is done by value sensitive design (Friedman and Kahn 2002).

Third, there is a kind of methods that focuses on existing technologies that have already to some extent been introduced in society. These methods offer ways of analysing technologies using particular normative frameworks, such as the ethical matrix (cf. Kaiser et al. 2007).

Both the societal initiatives and academic methods listed above fall under the general heading of responsible innovation. However, the field of responsible innovation is currently a fairly fragmented collection of projects and initiatives, and even corresponds to different academic discourses that do not sufficiently interact. For instance, the discourse on responsible innovation proper (cf. Owen et al. 2013; Von Schomberg 2013) focuses mostly on the *governance* of public, scientific research ("macro ethics"), while the literature in applied ethics that works with the research and innovation domain (cf. Friedman and Kahn 2002; Wright 2014) focuses predominantly on the *conduct* of practitioners and the negative impacts that their work might generate ("micro ethics"). At the same time, many thematic sub-domains of responsible innovation exist that focus on particular problem areas, privacy and safety being two prominent examples. We argue that this fragmentation is to a certain extent needed, but undesirable in its current state because it tends to generate initiatives for responsible innovation that have either a too narrow scale (e.g., purely focusing on professional ethics while neglecting the institutional, political dimension) or a too narrow scope (e.g., only focusing on privacy; thereby overlooking concerns of autonomy, safety, and so forth).

While acknowledging that no method will be truly comprehensive, in the sense of grasping the problematic of responsible innovation in its entirety or totality, the method proposed in this book at least aspires to this ideal. It traverses different domains, starting from the ethics of everyday technical practices in research and innovation contexts (e.g., coding a piece of software), to managing team dynamics of practitioners (e.g., enabling independent mediation by experts), to organising the institutional structures that govern technical practices (e.g., proposing structures for democratic conflict resolution). Through its appeal to virtue ethics, it also has a wide scope of concern, ranging from the individual virtues such as humility and courage, inter-personal virtues such as care and friendship, and institutional virtues such as justice and civility.

Finally, through its appeal to deontological concerns, it also links to the regulatory sphere in which compliance with professional codes of conduct and regulations and standards for the development of new technologies.

Towards a Hermeneutic Technology Ethics

This book develops *a hermeneutic ethics of technology* by connecting and bridging philosophical discourses on narrative theory, virtue ethics, and responsible innovation. We develop its central argument in different steps. We start from questioning the way in which the notion of technical practice has been conceptualised in contemporary philosophy of technology. As a response, we develop a notion of technical practice that draws from Ricoeur's work. Consequently, we link the resulting account of technical practice with a theory of technological mediation that draws from Ricoeur's narrative theory. Turning to ethics, we then formulate an account of narrative virtue ethics of technical practice, which lays out the way in which we can evaluate and prescribe technical practices at the personal, inter-personal, and institutional levels. Finally, we operationalise this ethical framework in the context of responsible innovation; outlining a method that captures how technical practices could be organised to be conducive to the good life, with and for others, in just institutions. These abovementioned elements translate into six upcoming chapters.

Chapter 2 surveys the theory of practice in the context of philosophy of technology. It first identifies some gaps in contemporary philosophy of technology. Existing theories of technological mediation pay too little attention to the linguistic, temporal, and social dimension of how technology shapes what we perceive and do. In line with this concern, the chapter observes that existing theories lack a comprehensive account of technical practice. It then surveys some authoritative accounts of practice (Aristotle, Marx, Bourdieu, and MacIntyre). Particular attention is paid to MacIntyre's conception of practice as a communal activity that is internally aimed at standards of excellence and it is argued that this can be used to understand *technology-in-practice* and, via this route, to evaluate technology by using a virtue ethics approach. For MacIntyre, practices and virtue have not only a social but also a narrative embeddedness—an element with which we will connect later in the book. Yet, the chapter also identifies some concerns arising out of MacIntyre's approach, which leads to a turn to other practice-oriented accounts of narrative theory; anticipating a turn to Ricoeur.

In Chap. 3, we aim to conceptualise a *hermeneutic* concept of technical practice. The unavoidable starting point for such an endeavour in philosophy of technology is Heidegger's engagement with Aristotelian praxis, both in his earlier work that outlined the question of Being and his later work on the Question Concerning Technology. We develop an original interpretation of Heidegger's notion of technical practice and argue that to address the enigma of authenticity it produces we have good reasons to turn to Ricoeur's work. Our argument shows how Heidegger's notion of "virtuous" practice is dependent on a concept of authenticity that prevents it from slipping into indeterminacy. However, in line with Ricoeur, we show that Heidegger is unsuccessful because of the arbitrary nature of his phenomenological solution and more radically because of the inability of hermeneutic phenomenology to solve the aporias of temporality. Consequently, we set out a theory of technical practice that is based on Ricoeur's theory of practice in *Oneself as Another* (1992). Our notion of technical practice operates on the methodological plane and captures three transitional stages: (1) from action to practice, (2) from practice to life plan, and (3) from life plan to narrative unity of life. As an illustration, this captures the idea that discrete acts like pressing different keys on a piano to play a melody relate to the practice of playing the piano, which relates to the life plan of being a (good) concert pianist, which is embedded in the notion of what a (good) pianist ought to be that is spread throughout a community.

Chapter 4 develops an account of technological mediation of technical practices that is inspired by Ricoeur's narrative theory as developed in *Time and Narrative* (Ricoeur 1983, 1985, 1988). It revolves around the notion of technical configuration. This notion operates on the methodological plane and captures three fundamental mediating structures: of (1) textuality, (2) temporality, and (3) distance. These structures are situated between the world of the human and the world of the technology. First, at the edge of the world of the human, the mediating structure is one of *textuality*—the extent to which a technology approximates the paradigm of the text and in that sense can be "written" and "read". This structure captures the active character of a technology, the extent to which it is attuned to the world of the human; and it captures the capacity of the human, the extent to which the human is attuned to the world of the technology. Second, in between the world of the human and the world of the technology is the mediating structure of *temporality*—the extent to which a technology mediates the chronological dimension of time. This structure captures the extent to which the technology enforces a certain temporal structure on human action; and it captures the extent to which a human is free in acting with the technology. Third, at the edge of the world of the technology, the mediating structure is one of *distance*—the extent to which a technology mediates events that are abstracted from the human world of action. This structure captures the extent to which a technology mediates a plot that stays proximate to the human world of action; and it captures the extent to which it mediates a plot that abstracts from the world of action, becoming increasingly inaccessible to human interpretation.

Chapter 5 develops the idea of a virtuous technical practice, as that which processes of responsible innovation ought to aspire to and can derive their standards from. To embed this discussion in the larger frame of the emerging field of virtue ethics of technology, we first engage in a detour that establishes the broad outlines of this field and presents the argument that it needs to engage with and develop an account of technological mediation to credibly aspire to be an established new framework in ethics of technology. As a response, we embed our narrative theory of technological mediation in the normative framework that conceptualises virtuous technical practice. Turning again to Ricoeur's work, notably his "Little Ethics" in *Oneself As Another* (1992), we combine a teleological and deontological perspective in constructing a hermeneutic ethics of technology. The notion of virtuous technical practice that grounds this ethics operates on the *normative plane* and traverses three stages: the good life (personal), with and for others (inter-personal), in just institutions (political). This structure allows us to pay attention to the professional ethics of engineers, the organisational ethics of teams of developers, and the institutional ethics of the institutions in which technologies are developed, such as universities, governments, and corporations.

Chapter 6 makes the final transition from normative ethics to the field of applied ethics and more specifically the field of responsible innovation. This step is important and urgent because it provides practitioners working in the context of research and innovation with practical guidelines for the design of tools, policies, and organisational structures to generate technical practices that are conducive of the good life. We present nine procedural steps that make up a new method for practising ethics in research and innovation contexts. To illustrate how each procedural step relates to concrete technical practices, a particular type of technology, namely personalisation technology, is presented as well as a concrete application of this technology to smart border controls. Subsequently, nine stages of responsible innovation are formulated, each containing distinct procedural steps. Our notion of virtuous practice design operates on the *prescriptive plane*, and captures three overarching phases, of (1) gathering and interpretation narratives (2) interpreting and technical practices, and (3) evaluating these practices according to the ethical aim, leading to different forms of prescription.

Chapter 7 concludes the journey of the book. It summarises the book's main contributions and provides suggestions for future investigations; most notably concerning a more politically oriented hermeneutic approach for responsible innovation. Overall, *Narrative and Technology Ethics* shows how humans and technologies *co-author* the technical practices that shape the world we live in. It makes a case for explaining these practices, in order to gain an enhanced understanding that will be needed for tackling today's problems in a virtuous manner.

References

Blok, V., & Lemmens, P. (2015). The Emerging Concept of Responsible Innovation. Three Reasons Why It Is Questionable and Calls for a Radical Transformation of the Concept of Innovation. In B. J. Koops, I. Oosterlaken, H. Romijn, T. Swierstra, & J. van den Hoven (Eds.), *Responsible Innovation*

2: Concepts, Approaches, and Applications (pp. 1–303). Heidelberg: Springer. https://doi.org/10.1007/978-3-319-17308-5.
Boenink, M., Swierstra, T., & Stemerding, D. (2010). Anticipating the Interaction between Technology and Morality: A Scenario Study of Experimenting with Humans in Bionanotechnology. *Studies in Ethics, Law, and Technology, 4*, 1. https://doi.org/10.2202/1941-6008.1098.
Bolsin, S., Faunce, T., & Oakley, J. (2005). Practical Virtue Ethics: Healthcare Whistleblowing and Portable Digital Technology. *Journal of Medical Ethics, 31*(10), 612–618. https://doi.org/10.1136/jme.2004.010603.
Brey, P. (2012). Anticipatory Ethics for Emerging Technologies. *NanoEthics, 6*(1), 1–13. https://doi.org/10.1007/s11569-012-0141-7.
Coeckelbergh, M. (2012). Care Robots, Virtual Virtue and the Best Possible Life. In P. Brey, A. Briggle, & E. Spence (Eds.), *The Good Life in a Technological Age*. New York: Routledge.
De Boer, B., Te Molder, H., & Verbeek, P. (2018). The Perspective of the Instruments : Mediating. *Foundations of Science, 23*, 739–755. https://doi.org/10.1007/s10699-018-9545-3.
Dorrestijn, S. (2011). Technical Mediation and Subjectivation: Tracing and Extending Foucault's Philosophy of Technology. *Philosophy & Technology, 25*(2), 221–241. https://doi.org/10.1007/s13347-011-0057-0.
Ess, C. (2009). *Digital Media Ethics*. Camebridge: Polity Press.
Feenberg, A. (1999). *Questioning Technology*. New York: Routledge.
Friedman, B., & Kahn, P. (2002). *Value Sensitive Design: Theory and Methods*. University of Washington CSE Technical Report, December, pp. 1–8. https://doi.org/10.1016/j.neuropharm.2007.08.009.
Frischmann, B., & Selinger, E. (2018). *Re-engineering Humanity*. Cambridge: Cambridge University Press.
Godin, Benoît. (2014). Ivention, Diffusion and Linear Models of Innovation: The Contribution of Anthropology to a Conceptual Framework. *Journal of Innovation Economics & Management, 3*(15), 11–37.
Graeber, D. (1996). Beads and Money: Notes Toward a Theory of Wealth and Power. *American Ethnologist, 23*(1), 4–24.
Hauser, J., Tellis, G. J., & Griffin, A. (2006). Research on Innovation: A Review and Agenda for Marketing Science. *Marketing Science, 25*(6), 687–717. https://doi.org/10.1287/mksc.1050.0144.
Ihde, D. (1979). *Technics and Praxis* (Vol. 33). Dordrecht: D. Reidel Publishing Company.

Jasanoff, S. (2015). Future Imperfect: Science, Technology and the Imaginations of Modernity. In S. Jasanoff & S. Kim (Eds.), *Sociotechnical Imaginaries and the Fabrication of Power* (Vol. 1, pp. 1–49). Chicago: The University of Chicago Press. https://doi.org/10.1017/CBO9781107415324.004.

Kaiser, M., Millar, K., Thorstensen, E., & Tomkins, S. (2007). Developing the Ethical Matrix as a Decision Support Framework: GM Fish as a Case Study. *Journal of Agricultural and Environmental Ethics, 20*(1), 65–80. https://doi.org/10.1007/s10806-006-9023-8.

Kajikawa, Y., Usui, O., Hakata, K., Yasunaga, Y., & Matsushima, K. (2008). Structure of Knowledge in the Science and Technology Roadmaps. *Technological Forecasting and Social Change, 75*(1), 1–11. https://doi.org/10.1016/j.techfore.2007.02.011.

Kaplan, D. M. (2006). Paul Ricoeur and the Philosophy of Technology. *Journal of French and Francophone Philosophy, 16*(1/2), 42–56.

Kudina, O., & Verbeek, P.-P. (2018). Ethics from Within: Google Glass, the Collingridge Dilemma, and the Mediated Value of Privacy. *Science, Technology, & Human Values, 44*, 291–314. https://doi.org/10.1177/0162243918793711.

Kurzweil, R. (1999). *The Age of Spiritual Machines*. New York: Penguin Group.

Latour, B. (1994). On Technical Mediation—Philosophy, Sociology, Genealogy. *Common Knowledge, 3*(2), 29–64. https://doi.org/10.1111/j.1365-294X.2010.04747.x.

Latour, B., & Porter, C. (1993). *We Have Never Been Modern*. Cambridge, MA: Harvard University Press.

Latour, B., & Venn, C. (2002). Morality and Technology. *Theory, Culture & Society, 19*(5–6), 247–260. https://doi.org/10.1177/026327602761899246.

MacIntyre, A. (2007). *After Virtue: A Study in Moral Theory* (3rd ed.). Notre Dame, IN: University of Notre Dame Press. https://doi.org/10.1017/CBO9781107415324.004.

March, S. (2014). Inside Steve Jobs Schools: Swapping Text-Books for Ipads. Retrieved from http://www.theguardian.com/teacher-network/teacher-blog/2014/oct/07/text-books-school-ipad-steve-jobs-classrooms.

Owen, R., Stilgoe, J., Macnaghten, P., Gorman, M., Fisher, E., & Guston, D. (2013). A Framework for Responsible Innovation. In R. Owen & J. Bessan (Eds.), *Responsible Innovation*. London: John Wiley & Sons, Ltd. https://doi.org/10.1063/1.1764494.

Pickering, A. (2010). *The Cybernetic Brain: Sketches of Another Future.* Chicago, IL: University of Chicago Press. https://doi.org/10.1108/03684921111133773.

Pinch, T. J., & Bijker, W. E. (1984). The Social Construction of Facts and Artifacts: or How the Sociology of Science and the Sociology of Technology Might Benefit Each Other. *Social Studies of Science, 14*(3), 221–232. https://doi.org/10.1177/030631284014003004.

Reijers, W., Wright, D., Brey, P., Weber, K., Rodrigues, R., O'Sullivan, D., & Gordijn, B. (2017). Methods for Practising Ethics in Research and Innovation: A Literature Review, Critical Analysis and Recommendations. *Science and Engineering Ethics, 24*, 1437–1481. https://doi.org/10.1007/s11948-017-9961-8.

Ricoeur, P. (1983). *Time and Narrative—Volume 1.* (K. McLaughlin & D. Pellauer, Eds., Vol. 91). Chicago, IL: The University of Chicago. https://doi.org/10.2307/1864383.

Ricoeur, P. (1985). *Time and Narrative—Volume 2.* (K. McLaughlin & D. Pellauer, Eds.). Chicago, IL: The University of Chicago.

Ricoeur, P. (1988). *Time and Narrative—Volume 3.* Chicago, IL: The University of Chicago.

Ricoeur, P. (1992). *Oneself as Another.* (K. Blamey, Ed.). Chicago, IL: University of Chicago Press.

Roberts, E. B. (2007). Managing Invention and Innovation. *Research Technology Management, 50*(1), 35–54. https://doi.org/10.1038/427679a.

Romele, A. (2019). *Digital Hermeneutics.* London: Routledge. https://doi.org/10.4324/9780429331893.

Romele, A., & Terrone, E. (2018). *Towards a Philosophy of Digital Media.* Cham: Palgrave Macmillan. https://doi.org/10.1016/S0031-9406(10)62747-2.

Simmel, G. (1900). *The Philosophy of Money.* (D. Frisby, Ed., 3rd ed.). New York: Routledge Classics, 1978.

Stiegler, B. (1998). *Technics and Time 1.* Stanford: Stanford University Press.

Teather, D. (2002). Lesbian Couple have Deaf Baby by Choice. Retrieved from http://www.guardian.co.uk/world/2002/apr/08/davidteather.

Vallor, S. (2016). *Technology and the Virtues: A Philosophical Guide to a Future Worth Wanting.* Oxford: Oxford University Press.

Verbeek, P. (2008). Obstetric Ultrasound and the Technological Mediation of Morality: A Postphenomenological Analysis. *Human Studies, 31*, 11–26. https://doi.org/10.1007/s10746-007-9079-0.

Verbeek, P.-P. (2005). *What Things Do; Philosophical Reflections on Technology, Agency, and Design*. Philadelphia, PA: Pennsylvania University Press.
Verbeek, P.-P. (2011). *Moralizing Technology: Understanding and Designing the Morality of Things*. Chicago, IL: University of Chicago Press.
Von Schomberg, R. (2013). A Vision of Responsible Innovation. In R. Owen, M. Heintz, & J. Bessant (Eds.), *Responsible Innovation*. London: John Wiley. https://doi.org/10.1002/9781118551424.ch3.
Wright, D. (2014). Ethical Impact Assessment. *Ethics, Science, Technology, and Engineering, 163*(c), 163–167. https://doi.org/10.1016/j.clsr.2011.11.007.Wright.

2

Praxis and Contemporary Philosophy of Technology

Today, philosophy and ethics of technologies is a burgeoning field of study. It has emerged from the far away, dark corners of dusty philosophy departments into the light of interdisciplinary research groups, "philosophy labs" and even commercial "ethics officers" for big Tech firms. This of course reflects the current of our time, especially in the wake of Covid19, in which each global problem at least in part seems attributable to technological innovation and its impacts. It has become more relevant than ever, and partly in response to this relevance, the field—at least in the English-speaking world—has largely made an "empirical turn"; turning away from heavy-duty theorising towards pragmatic analysis of the actual technological "things" that arise all over the place. This turn has been highly effective and produces a lot of valuable research. It has made us attentive to the manifold of technologies, their multistability, the many ways in which they are understood and used. As such, it has helped to build bridges between philosophy departments and science, technology, and society studies (STS), as well as media studies. Yet, the empirical turn has also, we believe, concealed some of the complexities of technical practice; the focal point of technology ethics. Turning it into an object of

W. Reijers, M. Coeckelbergh, *Narrative and Technology Ethics*,
https://doi.org/10.1007/978-3-030-60272-7_2

empirical analysis, technical practice has lost some of its ontological meaning (Zwier et al. 2016).

Placed in the context of the shortcomings of the empirical turn, this chapter revisits the notion of technical practice as focal point of philosophy of technology. As explained in the introductory chapter, we aim to develop a hermeneutic ethics of technology. For this purpose, we need a number of building blocks, including narrative theory, virtue ethics, and responsible innovation. Now, for preparing this work, we need at least the following two steps. To start, we need to argue why we need a hermeneutic ethics of technology at all in contemporary philosophy of technology: what are the problems or gaps in contemporary philosophy of technology to which we aim to offer a response? Consequently, in response to these problems and in order to prepare work on the (virtue) ethics part of our account, we need a philosophical understanding of technical practice, which starts with getting a better idea about what a "practice" is.

In what follows, we first identify central gaps in contemporary philosophy of technology: lack of attention to linguistic, social, and temporal dimension of technological mediation and absence of a theory of practice. We then review central theories of practice, with a focus on Alasdair MacIntyre's theory of practice in *After Virtue*. We interpret this theory, highlight the link with virtue and the narrative embeddedness of practice, and identify some problems in MacIntyre's account of practice. Third, the chapter concludes with the proposal of a "narrative turn" in philosophy and ethics of technology, which is central to the remainder of the book.

Gaps in the Empirical Turn

In order to develop a notion of *technical* practice, we need to learn from, and position our argument within, a discourse specialised in thinking about technology: philosophy of technology. Philosophy of technology has questioned the assumption that technology is a neutral tool, a mere means to our ends, and has shown that technology has many unintended effects. For example, people working in the phenomenological tradition have argued that technology harbours a particular way of thinking and relating to the world, and scholars in critical theory have pointed out how technology can lead to exploitation and support hegemonic social structures and relationships. Like much work in the humanities and social sciences in general, twentieth century philosophy of technology has been predominantly influenced by Heidegger and Marx, who each in their own way pointed to problematic features of modern society. Technology was seen in terms of "modern technology" (Heidegger 1977) or as a means of production which, under capitalism, becomes a tool of enslavement rather than emancipation (Marx 1977). The result was a rather bleak picture of technology as a driving force in promoting these problems: technology was seen as a tool of alienation, a way of thinking that seeks to dominate nature, and an instrument for oppression of the working class.

Contemporary philosophy of technology, in contrast, has been less concerned with the broader and structural effects of technology on human experience, existence, and society, and has been increasingly focusing on the technologies themselves—specific technologies. It has taken an "empirical turn" (Achterhuis 2001), meaning that it takes as its objects of philosophical reflection material artefacts (things) such as devices and machines, rather than text, discourse and narrative. Some philosophers of technology have in making this turn been inspired by science, technology, and society studies (STS), giving their work a strong empirical orientation and, inspired by Latour (1993), an orientation that focuses on symmetry between humans and non-humans. Moreover, contemporary philosophy of technology has sometimes taken an "engineering" approach which starts with an analysis of the nature of technology itself (Mitcham 1994, p. 62), connecting to the engineering sciences. The combined effects of these directions in philosophy of technology has been a progressive rejection of Heidegger's perspective, for example in the work of Don Ihde and Peter-Paul Verbeek, and more generally a turning away from the hermeneutic tradition. For example, following Ihde, Verbeek (2005) has rejected Heidegger's "pessimistic" views on technology at large (rather than concrete technologies). And Ihde (1998) has not only rejected Heidegger but has also taken distance from Paul Ricoeur, replacing textual hermeneutics with a hermeneutics of material things.

We argue that the change in focus that came with the empirical turn has led to the emergence of at least three gaps in contemporary philosophy of technology. The following reflection is not a comprehensive criticism, but rather a tentative diagnosis that will guide further efforts in the book. Not every gap is equally present in the main directions of the empirical turn (Postphenomenology, actor network theory (ANT), and Critical Theory of Technology). Yet, as will become clear in the next chapters, they all need to be addressed to come to a comprehensive notion of technical practice.

First, due to the focus on material artefacts, there has been insufficient attention to the role of *language*, including narrative (Coeckelbergh 2017a, b). Technological artefacts are assumed to mediate the relation between us and the world, but without any role assigned to language; the words we use and the narratives we recount to ground our understanding of technology. For instance, when a device with voice interface mediates between me and the world, this is a technological artefact but any account of what it is and does crucially needs a reference to language. Or consider an android robot that evokes fear as a response: to understand this, we have to refer to prefigured understanding, shaped by narratives about technology such as Mary Shelley's *Frankenstein*, a story which is usually interpreted as a warning for more intelligent and human-like technology running out of control (or, a story about a human creator abandoning his creation). Philosophy of technology also pays little attention to the symbolic qualities of media and technologies. Consider money, for instance, which is not merely—and not necessarily—a "material artefact" but a tool that mediates relations between humans in a way that can only be explained by reference to its symbolic function. Perhaps contemporary philosophy of technology puts so much emphasis on the material dimension of technologies, that the technologies of writing and text—and how they continue to mediate and shape our knowledge and experience—receive considerably less attention. In Ihde's posthermeneutics of things and in Verbeek's account of what things do, there is no longer space for language—not even as a mediator. Philosophical hermeneutics, in contrast, has a lot to say about that. Yet, by taking an empirical turn, there was no longer any. interest in narrative theory or language. As such, the baby has been thrown out with the bathwater.

Second, theories of technology after the empirical turn do not exhibit much sensitivity to the *temporal* dimension of the use of technology, if they consider it at all. The postphenomenological analysis is indifferent regarding time: it does not consider that it is made at one point in time—a momentary intervention of technology—and human subjectivity in relation to the technology is understood in a rather a-temporal way. Similarly, ANT and Latour's non-modern conception of the social largely disregard temporality. As Romele (2020) has put it, ANT is a kind of "flatland"; which also involves the flattening out of the temporal dimension of actor networks and their connection with human subjectivity. We do not hear about how humans experience time, how their technology use is part of a life they lead, how they understand that life in a temporal way. Also, Critical Theory of Technology, though being attentive to the "concretisation" of technical practices, does not thematise their temporality (Feenberg 1999). And yet, one of the fundamental insights of Heidegger's hermeneutical method has been that we need to approach the meaning of praxis through an interpretation of temporality. Turning away from Heidegger, it seems, has also caused a turning away from the question of time.

Third, in particular Postphenomenology and mediation theory tend to focus on an individual "I" that relates to the world via technology. This has been mostly inherited from the subjectivism in Husserl's phenomenological method, and its focus on pure perception of an individual ego. The *social* aspect of how technology mediates and shapes our experience and action is under-theorised, if not absent. Ihde's Postphenomenology does not pay much attention to the social and instead focuses on how technology mediates human-world relations (Ihde 1990). To use Van den Eede's phrase: Ihde and Verbeek say a lot about how technology is in between I and the world; what is missing is how technology is "in between us" (Van den Eede 2011, p. 157). ANT as well as Critical Theory of Technology have been more concerned with the social and political aspects of technology. However, they have not adequately managed to bridge the gap between the 'micro' of individual practices and the 'macro' of societal structures. This bridging, as we shall see, is offered by philosophical hermeneuticists, most notably by Ricoeur.

Keeping in mind that our book aims to develop an *ethics* of technology, we argue for the need for an adequate account of technical practice that responds to the gaps identified: *language, temporality*, and *the social*. We will elaborate these aspects in the next chapters. Before arriving at a notion of a technical practice, however, we explore what key philosophers have meant with *practice*. Let us take a brief look at the philosophy of practice, with a focus on the work of MacIntyre.

Revisiting Practice

The notion of practice, derived from the Aristotelian term *praxis*, has been discussed throughout the history of philosophy. Here, we provide an overview of this discourse on practice, focusing on some of the most influential account, to lay the grounds for an understanding of *technical practice*.

According to Aristotle, there are three kinds of human activities: *theoria* (thinking, theory), *poiesis* (bringing forth; skilful making of things), and *praxis* (doing). Praxis can take the form of political action, where it becomes discourse between free men. In the *Nicomachean Ethics*, Aristotle connects praxis with deliberation (the origin of action is choice, 1139a2) and gives it an ethical meaning: it is an activity done for the sake of the activity itself—an ethical activity which is also part of activity in the *polis*. Drawing on Aristotle's distinctions and focusing on praxis as political action, Arendt (1958) argues that Western philosophy has often focused on the contemplative life at the cost of neglecting praxis and the active life, in particular political action. Yet, *praxis* can also refer to action in general, as in the *Poetics*, where Aristotle defines tragedy as the imitation (*mimesis*) of action (1449b24).

For Aristotle, praxis *can* already have a political dimension or is at least embedded in a conception of the political life. But it is in Karl Marx's work that practice takes on a stronger social and political meaning. Marx, borrowing the concept of praxis from his fellow nineteenth century thinkers of the left Hegelian movement, who also stress praxis as opposed to contemplation. For Marx, praxis is the conscious, free, productive, and creative activity through which humans shape their history and themselves. Revolutionary practice, then, can change the world. In the *Theses on Feuerbach* (1976), Marx argues against idealism and the distain for practice in Western philosophy. Truth is a practical question and all social life is practical. Through revolutionary practice we can change our circumstances. (Marx 1976, p. 4). His 11th thesis famously reads: "The philosophers have only interpreted the world in various ways; the point is to change it" (Marx 1976, p. 5). Praxis thus has a social significance.

Pierre Bourdieu, notably influenced by Heidegger, Wittgenstein, Merleau-Ponty, and Marx, also gives a social meaning to practice, but in a different way. In his practice theory, partly developed in response to structuralism, the concept of "habitus" is central. As a sociologist he was interested in conceptualising the relation between the individuals and their social environment, between individual agency and social structure. The concept of habitus delivers this: situated at the meeting point of social structure and individual minds-bodies, it captures a lasting internalisation of the social order in human bodily existence. It connects social order with what we may call bodily comportment (Bourdieu uses the term *hexis*, which derives from the Greek verb of "to have"); which is at the same time social and ethical comportment. The term habitus does not just mean disposition in an abstract sense; the social structure is *incorporated*. It is about habits in thinking but also about intuitions, feelings, skills, and ways of bodily doing. One holds and uses one's body in certain ways, uses one's voice in certain ways, and so forth; and this is socially shaped. In *Distinction* (1984), Bourdieu writes that social class, for instance, is not just a matter of one's position in relations of production, as Marxists argue, but is also defined by a "class habitus" associated with that position, such as judgments of taste. Interestingly (if we keep in mind Aristotle), habitus is acquired through imitation (*mimesis*). In this way, society and culture shape the body and minds of individuals, which in turn enables these individuals to find new solutions to new situations. While "agents are possessed by their habitus more than they possess it" (Bourdieu 1977, p. 18), Bourdieu keeps the door to social change open. Habitus is a product of history but also produces new and "more history", new "individual and collective practices" (Bourdieu 1990, p. 54).

This is done without a conscious aiming. As Bourdieu puts it in *The Logic of Practice*, habitus is defined as:

> systems of durable, transposable dispositions, structured structures predisposed to function as structuring structures, that is, as principles which generate and organise practices and representations that can be objectively adapted to their outcomes without presupposing a conscious aiming at ends or an express mastery of the operations necessary in order to attain them. (Bourdieu 1990, p. 53)

Interestingly, this definition follows a comment on Marx's *Theses on Feuerbach*: Bourdieu suggests he takes distance from Marx by claiming that we should situate knowledge within "real activity as such", that is, "in the practical relation to the world", and proposes nothing less than a "return to practice". (Bourdieu 1990, p. 52)

Habitus thereby enables social organisation in a less visible, less explicit way than Marx envisioned. Bourdieu writes about the conditions associated with a particular class: "objectively 'regulated' and 'regular' without being in any way the product of obedience to rules, they can be collectively orchestrated without being the product of the organising action of a conductor" (Bourdieu 1990, p. 53). This way of approaching social order and power resonates with Michel Foucault. Foucault's concepts of discipline and technologies of the self can be seen as related concepts: discipline is power impressed on the body and techniques of the self are ways how the social order affects "bodies and souls, thoughts, conduct, and way of being" (Foucault 1988, p. 18). One could also consider Marcel Mauss's "body techniques" or "techniques of the body" (1973). Mauss uses the term to describe boy actions that embody aspects of a society or culture. In every society or culture people know how to use their bodies in a way that may differ from other societies and cultures. For example, people may walk in different ways or use their hands in different ways. This again constitutes a way to relate individuals and their bodies to wider social and cultural environments.

Bodies are not understood in terms of objective or symbolic facts, as the objectivism of French structuralism has it according to Bourdieu (cf., 1977, p. 23); instead, he shifts the emphasis from symbols, consciousness, and language to agent's practices, the role of the body in these practices, and the work necessary to maintain them. While he does not offer a neatly delineated definition of practice, Bourdieu offers examples from anthropological fieldwork and offers an interesting metaphor: when he opposes the (structuralist) anthropologists' description of logical relationships to practical relationships, he explains the latter as follows:

> practical because continuously practiced, kept up, and cultivated—in the same way as the geometrical space of a map, an imaginary representation of all theoretically possible roads and routes, is opposed to the network of beaten tracks, of paths made ever more practicable by constant use. (Bourdieu 1977, pp. 37–38)

Hence, practice is about use, and use has a temporal meaning. The anthropologist looks at practices from a distance and makes it into a representation and the structuralist anthropologist describes objective relations, grasping practice from the outside. But this standpoint has its limits. Bourdieu does not argue against conceiving of social structures as such, but stresses that they are always related to practices. Habitus thus negotiates between objective social structures and practices, with the balance in favour of the latter, but always connecting both. With the notion of habitus, Bourdieu tried to overcome theory-practice dualism.

With Marx and Bourdieu, praxis thus gets a more social and bodily, or performative meaning than in Aristotle. This understanding of practice also offers an interesting way to connect the symbolical to the empirical, and hence can also be used in response to the problems raised by the empirical turn. This is not limited to using the concept of habitus. Drawing on Bourdieu's sociology of the symbolic forms, Romele has recently argued that Bourdieu's approach may benefit philosophy of technology if we use it to argue that "technologies are always already entangled in social (and cultural) dynamics of classification, separation, and eventual exclusion and discrimination"—thus reintroducing a transcendental dimension without reverting to dogmatism (Romele 2020, p. 2). However, it remains controversial how much determinism Bourdieu's theoretical framework implies: are individuals and their actions and bodily comportments, through habitus, not the *mere* expression of social structure?

Unlike Bourdieu, MacIntyre stays closer to Aristotle, and does not adopt the Marxist (or other, sociological) interest in social order.

MacIntyre also loosens the connection between practice and the body (and the attention to subconscious action), and at the same time suggests a narrative interpretation of praxis. Moreover, in contrast to Bourdieu, MacIntyre offers an *ethics* and enables us to say more about *virtuous* technical practice. In *After Virtue* (2007), MacIntyre accounts of human good in social and narrative terms: "in terms of practices, traditions, and the narrative unity of human lives" (MacIntyre 2007, p. xi). The notion of practice plays an important role in this. Inspired by Aristotle and learning from history and anthropology, MacIntyre argues that ethical inquiry is situated in the contexts of practice. He criticises the modern obsession with autonomy, the modern views of moral agency that disconnect it from any social roles and contexts, and the modern belief that reason in the realm of practice can only speak of means, not ends. Virtues are generated in everyday life, through "the engagement by plain persons in a variety of practices, including those of making and sustaining families and households, schools, clinics, and local forms of political community" (MacIntyre 2007, p. xv).

MacIntyre's account of ethics in terms of practices raises the problem according to which moral standard these practices are to be evaluated. His Aristotelian (and Thomist) answer is that there is an "end" to which all these practices are directed. And this end is linked to the question: "what sort of person am I to become?"—an answer to which is again to be found in practice in each human life (MacIntyre 2007, p. 118). The idea is that once we have an understanding of the virtues, acquired by engagement in and with our practices and communities, this gives us guidance concerning right or just (MacIntyre 2007, p. 133). It enables us to evaluate our practices. This reasoning seems circular, but—one could argue in defence of MacIntyre—this is a "healthy" circularity once we conceptualise the relation between the person and community as a hermeneutic one: being engaged in a practice and being part of a community gives us an understanding of ethics and morality, which then can be used to interpret and evaluate particular practices. While MacIntyre's answer is not entirely satisfactory, as we explain later, and while this account of good is anthropocentric, it is an interesting account of practice as related to the human good, which offers an effective criticism of modern approaches to morality that see good as entirely divorced from practices and communities. But what are practices? And what exactly is their relation to good and virtue? This is how MacIntyre defines "practice":

> By a practice I am going to mean any coherent and complex form of socially established cooperative human activity through which goods internal to that form of activity are realised in the course of trying to achieve those standards of excellence, which are appropriate to, and partially definitive of, that form of activity, with the result that human powers to achieve excellence, and human conceptions of the ends and goods involved, are systematically extended. (MacIntyre 2007, p. 187)

This definition already includes some claims about the link between practice and good, and to understand it we have to turn to MacIntyre's Aristotelian conception of ethics. For MacIntyre, every practice aims at some good. Following Aristotle, he claims that the good is linked to the nature of human beings, which gives them certain aims and goals "such that they move by nature towards a specific *telos*" (MacIntyre 2007, p. 148). Aristotle defines this end as *eudaimonia*, often translated as happiness. The virtues are then "those qualities the possession of which will enable an individual to achieve *eudaimonia*" (MacIntyre 2007, p. 148). Aristotle leaves it open what this is: we have to find out by living our lives. It is to be found in the exercise of the virtues, which is not a mere means to the end of the good. MacIntyre writes: "For what constitutes the good for man is a complete human life at its best, and the exercise of the virtues is a necessary and central part of such a life, not a mere preparatory exercise to secure such a life" (MacIntyre 2007, p. 149). Particular practices, then, enable us to exercise and exhibit virtues that have social significance. MacIntyre mentions arts, sciences, games, and politics. These practices have goods that are internal to them, which means that the practice itself exhausts the good that it fulfils.

Narrative Embeddedness of Practice

Practices involved in the acquisition of virtue thus have an important social and public dimension. According to MacIntyre, goods can only be achieved "by subordinating ourselves within the practice in our relationship to other practitioners. We have to learn to recognise what is due to whom", try to improve ourselves, and so on (MacIntyre 2007, p. 191). We learn for example courage or justice in relation to others, indeed by *practicing* it with others and towards others. These others are not only present others but also historical others. There is a tradition, for example a tradition of painting. MacIntyre writes: "To enter into a practice is to enter into a relationship not only with its contemporary practitioners, but also with those who have preceded us in the practice, particularly those whose achievements extended the reach of the practice to its present point" (MacIntyre 2007, p. 194). For example, a contemporary painter will learn from, and have to respond to, a particular painting tradition. This is not only true about learning technical skills, but also about learning the virtues, according to MacIntyre. Without virtues, practices would be corrupted by institutions, which are directed towards external goods such as power, status, and monetary gain. Vice corrupts institutions and endangers the achievement of internal goods.

MacIntyre admits that an inherent morality of virtue is not enough; practices also have to meet moral standards (MacIntyre 2007, p. 200). However, within his framework it is not clear where those (external) moral standards would come from, if not from the practices themselves. This is again the problem of circularity pointed out earlier. How does MacIntyre deal with this problem?

First, his Aristotelian and hermeneutic response is to look at the life of an individual as a whole, a life conceived as a unity. The idea is that there is an end (*telos*) that transcends the goods of practices "by constituting the good of a whole human life, the good of a human life conceived as a unity" (MacIntyre 2007, p. 203). This understanding of virtue and good is interesting, since it is connected to narrativity. We will say more about this later.

Second, MacIntyre's theory of action is social and relational, and—sometimes explicitly, sometimes implicitly—narrative. He speaks of "episodes" and "settings". Agents have intentions, but these intentions are only intelligible in a social setting such as a practice or another human *milieu*. Moreover, that setting has a history and individual agents have to be situated within that history (MacIntyre 2007, p. 206). This provides us with a temporal aspect. Human actions are characterised within narrative history. Now, analytic philosophers would also say that there are sequences of human actions. But, MacIntyre argues, an action is always embedded in a context: it is an event in a history. Actions are only intelligible within the whole of the sequence, and within a context. Moreover, human lives as wholes have a narrative structure and humans are narrative beings: "man is in his actions and practice, as well as in his fictions, essentially a story-telling animal". In order to answer the question what to do, I have to answer the question "Of what story or stories do I find myself a part?" (MacIntyre 2007, p. 216). Our lives are stories, and if we seek the good life, we are on a "quest for the good". Virtues sustain communities "in which men and women can seek for the good together" (MacIntyre 2007, p. 219). And again, this refers to a historical process or dynamic. We seek the good within a particular practice and community, which have histories and traditions. This is a form of *narrative embedding*. The good we seek as persons is not only linked to present practices but also to traditions, which—at least if they are living traditions—provide a context and narrative for those practices and are at the same time sustained by these practices. Living traditions "continue a not-yet-completed narrative" (MacIntyre 2007, p. 223).

MacIntyre's conception of good is thus closely linked to the notion of practice and "the narrative unity of a human life and of a moral tradition" (MacIntyre 2007, p. 258). Narrativity is a dimension he develops throughout his elaboration of his account of virtuous practice. After the first stage of that elaboration (virtue is needed to achieve goods internal to practices), narrativity comes in when virtues are defined as qualities contributing to the good life considered as a narrative whole and when virtue is seen as part of the pursuit of human good within an "ongoing social tradition" (MacIntyre 2007, p. 273). Virtue (and more generally moral agency) is not just defined in social and practical terms; it is also embedded in a temporal order, in particular in a narrative context. In order to be a virtue at all, a quality needs to be operative in these narrative stages as well. In this way, ethics is not only a matter of the right or wrong actions of individual (atomistic action) but is integrated in questions concerning the story of our lives and the stories of our community and tradition. If I want to be virtuous, I have to ask: "What is the best kind of life for a human being like me to lead?" and, in the light of my tradition, how can I integrate this in the patterns of my tradition and contribute to that tradition, which is a quest for "*the* good and *the* best"? (MacIntyre 2007, p. 275)

Objections to MacIntyre

Our traversal of the philosophy of practice, and notably the account of practice in MacIntyre, has given us important touchpoints to address the gaps in philosophy of technology after the empirical turn. First, we saw that this philosophy of practice deals with the linguistic and symbolical mediation of practice, most notably through the concept of narrative. Second, the philosophy of practice assigns an unmistakable social dimension to practices and connect the micro level of individual practices with the macro level of social structures. Third, we find that temporality and historicity is a recurrent theme in the philosophy of practice, because practices have a historical significance in moral communities, as well as a temporal significance in denoting activities that have a certain duration. Given these initial findings, MacIntyre's theory of practice would be a very promising starting point for initiating a dialogue with contemporary philosophy of technology and gaining a better understanding of what is meant with technical practice.

However, MacIntyre's theory has certain drawbacks; three of which are central given our purposes. First, MacIntyre posits an opaque, yet also idealised notion of practice. His notion is opaque because it includes criteria of *complexity, coherence, cooperative activity* and *systematic extension* that lack qualification. That is, his theory of practice does not admit for questions such as: "Which activities count as complex and which as simple?", "What does it mean for an activity to be coherent?", "why would only cooperative activities count as practices?", and "what does it mean for internal goods to be systematically extended?" One can for instance question whether "playing the violin", which MacIntyre designates as a practice, is a cooperative activity because one can play the violin exclusively by oneself. Perhaps playing the violin is "social" in a different sense, but this is not clear in his account. MacIntyre's very extensive notion of practice thereby creates a problematic degree of conceptual opacity.

At the same time, MacIntyre's notion of practice is idealised (cf. Carr 2003, p. 256). On the one hand, it seems to surpass Aristotle's notion of praxis in its narrow sense—of being the kind of action that allows for the cultivation of the virtues—, by admitting for practices such as playing chess, the violin or the piano, which Aristotle would have arguably considered as belonging to *technê*, to the crafts or arts that serve an instrumental purpose and are not concerned with the virtues.[1] On the other hand, however, MacIntyre counts *only* those activities as practices that realise goods that are internal to the particular type of activity. In this regard, he idealises Aristotle's notion of practice by conceding that for an activity to be regarded as a practice, it necessarily needs to support the cultivation of the agent's virtue. But for Aristotle, any practice is still praxis, even if the goods internal to it are not aimed for or not reached. The activity can also lead to a deficiency or an excess of a feeling or thought; for instance, when the agent's wish that constitutes the practice is the *wrong* one. For Aristotle, therefore, a warrior engaging in battle and charging hesitantly and thereby putting his comrades in danger still engages in a practice, though he can also be called a coward (showing a lack of virtue). For MacIntyre, however, this would disqualify the very activity from being a practice, leading to a much narrower and more idealised notion of practice than the one proposed by Aristotle. The major problem of this narrow conception of practice is that it creates a dichotomy without proper qualification. For instance, a violin player might at first be motivated by her parents to play the violin (e.g., realising the

[1] In this choice, however, we will follow MacIntyre because we reject Aristotle's strictly instrumental notion of *technê*. As we will argue in Chap. 6, instrumentality should be considered as a matter of degree, rather than a matter of different kinds that somehow allow for "pure forms" (e.g., a craft being "purely instrumental).

external good of honour) and gradually move to be motivated according to the internal goods (e.g., mastery) of the activity. In MacIntyre's framework, "playing the violin" would thereby move from being a non-practice to a practice. However, it is not at all clear at what point we can designate the activity as a "genuine" practice in its idealised sense. This problem becomes increasingly pressing when activities become more complex, notably in technologically mediated environments (e.g., when considering practices of stock market trading or playing video games).

Second, MacIntyre uses a restricted notion of narrative, being only tied to a person's history in the sense of stories "told in the thick of everyday life" (Ricoeur 1992, p. 158). As such, MacIntyre argues that we make sense of our practices by relating them to our recounted past experiences and the past experiences of others. However, by taking this idea of personal or common historical narrative as unproblematic, MacIntyre fails to account for the difference between historical and fictional narrative and how these both narrative modes inform our practices. As Ricoeur concedes (Ricoeur 1992, p. 158), the reinterpretation or refiguration of our practices through fiction produces a thorny problem for MacIntyre's theory. The problem that MacIntyre faces is that, because of its productive qualities, literary fiction can configure novel interpretations of practices and the standards of excellence through which the internal goods to these practices are realised that sometimes are in discord with a moral community. A historical example would be *Uncle Tom's Cabin*: a fictional narrative that broke with the ideas of virtuous practices of the moral community of the time (of slave owning and management) that it addressed. As Carr argues, MacIntyre's theory opens up the possibility of a problematic type of moral relativism (Carr 2003, p. 257), because the criteria for a practice to be desirable or not can only be found in the historical narrative surrounding an agent and her moral community. However, this moral community can have conceptions of practices, their internal goods, and standards of excellence that are quite clearly morally wrong (as illustrated by example of slavery). Uncritically importing MacIntyre's theory of practice would thus create a vulnerability to this objection.

Third, MacIntyre fails to properly integrate or relate the three major components of his theory of practice: practice, narrative unity and moral community. For practices, he argues that they are made intelligible through narrative structures, but he does not provide an account of *how* this happens. For narrative structures, similarly, he argues that such structures fit within a moral community that has a certain conception of the good life but does not provide an account of *how* they fit. As Devine argues, "there is a gap in MacIntyre's account between the concept of a practice and the concept of a tradition" (Devine 2013, p. 115): it is unclear how heterogeneous sets of practices (e.g., having a conversation, playing an instrument, conducting political debates) combine through narratives to form a tradition. The connection between the different parts of MacIntyre's theory will need to be made explicit in order to move beyond anecdotal evidence. For instance, *that* we commonly explain what we do when engaged in a practice by telling a narrative does not by itself offer sufficient grounds for assuming that therefore practice and narrative are necessarily connected and should be integrated in a single theory of practice.

A Narrative Turn

We now have an initial understanding of what practice is about. But this does not give us an account of *technical* practice. How to bring in technology? What is still missing is a theory of technical practice that incorporates the notion of technical mediation—accounting for the role that technological artefacts and systems play in mediating a technical practice.

One way to deal with the missing aspect of technological mediation would be to drop mediation theory, steer away from the notion of practice and replace it by an entirely different concept. Drawing on Wittgenstein and philosophy of language more broadly, Coeckelbergh has proposed the notions of "technology games", "performance" and related performance metaphors to conceptualise how technological use is always related to wider social and culture wholes (Coeckelbergh 2017a, b, 2019; Coeckelbergh and Funk 2018). Instead of focusing on mediation by technological artefacts, as postphenomenology and mediation theory do, "technology games" and "technoperformances" shift the attention the human users and their bodies, relationships, language, societies, and cultures, and enable a more social and holistic approach. This direction responds to the gaps identified in the beginning of this chapter, could also serve as a basis for an ethics of technology, and—via the performance metaphors—enables to bring in narrativity theory. In principle, through its focus on use, one could also link it to practice and hence to the virtue ethics tradition including MacIntyre. However, this approach largely gives up mediation theory and does not fully engage with the *hermeneutic* tradition in philosophy.

Another way to bring together practice and technological mediation is to keep the notion of technological mediation and more directly connect with the hermeneutic tradition, by offering narrative theory as a way to address the challenges of conceptualising technical practice. Looking back at our review of the philosophy of practice, we see that narrative plays an important role as concept of mediation. That is, narratives mediate the temporality of practices, connect them to life plans and historical traditions, and turn individual practices into social activities. Considering narrative as the key to bringing practice and technological mediation together is the route we choose in this book. Hence, we propose a "narrative turn" in philosophy and ethics of technology.

But what is "narrative", precisely? Jerome Bruner characterises narrative as "a conventional form, transmitted culturally and constrained by each individual's level of mastery and by his conglomerate of prosthetic devices, colleagues, and mentors" (Bruner 1991, p. 5). Alternatively, narrative can be characterised by its mediating role (cf. Ricoeur 1983)—between a human being and her experiences and understanding of the world. Plainly said, deriving from the Aristotelian tradition, a narrative like a story in a book is defined by the way in which it mediates the experience and understanding of the reader's social world. This mediating role can be captured by the concept of *emplotment*, which designates the capacity of a narrative to organise heterogeneous elements (characters and events) in a meaningful synthesis that we commonly designate as the plot. As such, we characterise narrative as a mediation of human experience and understanding of the social world through the process of emplotment, which is the organisation of heterogeneous elements in a meaningful synthesis. The paradigmatic model for a narrative (its empirical correlate) is the model of the text, which is used as the basis for understanding how narratives mediate our experience and understanding of the world. However, narrative is not exclusively bounded to actual texts because it also mediates utterances (e.g., telling a story), performances (e.g., a tragedy, a comedy) and—as we will argue—technical practices (e.g., using a computer interface, driving a car).

There are various philosophical views on how the notion of narrative can contribute to our understanding of the social world and on the way in which they shape social reality. Some scholars consider narrative as an instrumental cognitive ability or linguistic tool, whereas others consider it as an ontological category connected to the way humans are in the world (Meretoja 2014, p. 89) or understand human life itself as having a narrative character (Macintyre 2007, p. 114). Another theoretical division with regard to the role of narrative exists between an empiricist tradition that denounces narrative as a fundamental philosophical concept (c.f., Strawson 2004) and the hermeneutic tradition that instead rejects the idea of experience unmediated by narratives. The latter tradition holds that all representations of the human social world are mediated by human-linguistic interpretation (Taylor 1971, p. 4), that subjectivity is always mediated by language, "by 'signs, symbols and texts'" (Meretoja 2014, p. 96). Thus, rather than viewing narrative merely as an instrumental or discursive tool that shapes the design process of technologies, it should be understood as a fundamental ontological aspect of human social reality.

Beyond philosophy, multiple scholars have shown how narrative ontology can improve our understanding of concrete aspects of our social world. As such, the social sciences have witnessed "narrative turns" since the 1980s (Hyvarinen 2010), which denote different resurgences of interest in narrative theory for purposes of conceptualisation, historical analysis, and methodology. For instance, Bruner explains how we can increase our understanding of human psychology, as embedded in a cultural context, by looking at how a "text affects the reader" (Bruner 1991, p. 4). Accordingly, he claims that narratives can "make events" and even "make history" (Bruner 1991, p. 42). In a similar vein, Czarniawsma employs a methodology for organisation studies that allows for an understanding of the "reflexive nature" of the human condition as the basis for collective action (Czarniawsma 1998, p. 77). She understands an organisation as a story, understood as a social construct that is shaped by human interaction and interpretation through narratives. Gotham and Staples show how the significance of narrative goes beyond our understanding of history and can help analyse "human agency in processual, action-oriented ways" (Gotham and Staples 1996, p. 492). They argue that a narrative understanding of our social world would improve sociological inquiries. These and many other scholars have shown how a narrative ontology can be fruitfully employed in studies of different aspects of our social world. We will build on this idea, by exploring how narrative theory, focusing on the work of Ricoeur, can assist us in studying the role of technologies in our lifeworld.

In taking a narrative turn, the challenge for us is the following. First, in response to mediation theory, we aim to use narrative theory as a theory of technological mediation that conceptualises how our understanding and actions are mediated by external things (texts, but also other technological objects). Thus, we do not reject the conceptualisation of technology in terms of material artefacts, but we add that texts also mediate, put technology in the context of a practice, and offer a narrative theory of mediation and practice, which (among other things) adds the dimensions of language (in the form of narrative), temporality, and the social to existing accounts of mediation as developed by Ihde and Verbeek. We also position our approach within an overview of use of narrative theory in other fields.

Second, in order to develop our theory and ethics of technology, we draw on a robust *narrative* theory in philosophy. For this, we will be mainly guided by Ricoeur, and thereby go beyond earlier attempts to use Ricoeur for understanding technology (cf. Kaplan 2006). As said in the first chapter, Kaplan used Ricoeur's narrative approach to argue that we make sense of technology and our lives in narrative ways. But we go beyond this initial diagnosis and argue that technology also configures our actions and life narratives. In terms of approach, we base our analysis on a thorough interpretation of Ricoeur's narrative theory. In the next chapters, we will show that Ricoeur's narrative theory is not only a promising touchpoint for thinking about technical practice, but also helps us to develop a narrative ethics of technology. Our challenge is to interpret and use Ricoeur's narrative theory in a way that (1) enables us to merge it with an account of practice and (2) use that for crafting a narrative ethics of technology, understood as a narrative ethics of technical practice. As we will discover in the following chapters, Ricoeur in fact took us a long way already in drawing these connections.

Conclusion

This chapter placed our investigation in the larger context of philosophy and ethics of technology and the philosophy of practice. First, we identified some gaps in contemporary philosophy of technology after the empirical turn, in particular Postphenomenology and mediation theory. These concerned a neglect of language, temporality, and the social in discussions of technical practice. Second, we revisited the philosophy of practice, and discovered how it deals with the gaps identified. In particular, we discussed MacIntyre's notion of practice as a promising point of departure, but also considered some of its shortcomings. Finally, we concluded that ethics and philosophy of technology is in need of a "narrative turn", in order to link the philosophy of practice with an account of technological mediation.

From the foregoing discussions, it follows that we need an alternative approach to understand how language, collective intentionality and technology are intertwined in the normative shaping of our lifeworld. Drawing from the idea that a "competition" between narratives can result in technology change, as Dawson and Buchanan (2005) suggest, we propose to re-frame the problem they present by asking: how do humans and technologies *co-author* narrative structures that have the potential to transform our understanding of our technologically mediated social reality? An approach that follows this question might answer to the concern raised by Feenberg (2009, p. 228) about postphenomenological theories of technology and theories in STS, namely that they provide for a convincing relational ontology but insufficiently offer a corresponding hermeneutic theory of meaning (Feenberg 2009, p. 228). At the same time, we have to resist ideas of technological determinism such as the one proposed by Winner (1980), because our framework should explicitly acknowledge the agency of both humans and technologies in the mediation of social reality.

These concerns cause us to turn to narrative theory, and notably to the work of Ricoeur. In the next chapter (Chap. 3), we will problematise the concept of technical practice in order to provide a more elaborate rationale for our choice to use Ricoeur; and subsequently explicate his theory of technical practice.

References

Achterhuis, H. (2001). *American Philosophy of Technology: The Empirical Turn.* Bloomington, IN: Indiana University Press.

Arendt, H. (1958). *The Human Condition.* Chicago, IL: Chicago University Press.

Bourdieu, P. (1977). *Outline of a Theory of Practice* (R. Nice, Trans.). Cambridge: Cambridge University Press.

Bourdieu, P. (1984). *Distinction: A Social Critique of the Judgement of Taste* (R. Nice, Trans.). Cambridge, MA: Harvard University Press.

Bourdieu, P. (1990). *The Logic of Practice* (R. Nice, Trans.). Cambridge: Polity Press.

Bruner, J. (1991). The Narrative Construction of Reality. *Critical Inquiry, 18*(1), 1. https://doi.org/10.1086/448619.

Carr, D. (2003). Rival Conceptions of Practice in Education and Teaching. *Journal of Philosophy of Education, 37*(2), 253–266. https://doi.org/10.1111/1467-9752.00324.

Coeckelbergh, M. (2017a). Language and Technology: Maps, Bridges, and Pathways. *AI & Society, 3*, 175–189.

Coeckelbergh, M. (2017b). *Using Words and Things: Language and Philosophy of Technology*. New York, NY: Routledge.

Coeckelbergh, M. (2019). *Moved by Machines: Performance Metaphors and Philosophy of Technology*. New York, NY: Routledge.

Coeckelbergh, M., & Funk, M. (2018). Wittgenstein as a Philosopher of Technology: Tool Use, Forms of Life, Technique, and a Transcendental Argument. *Human Studies, 41*(2), 165–191.

Czarniawsma, B. (1998). *A Narrative Approach to Organization Studies*. London: SAGE Publications.

Dawson, P., & Buchanan, D. (2005). The Way It Really Happened: Competing Narratives in the Political Process of Technological Change. *Human Relations, 58*(7), 845–865. https://doi.org/10.1177/0018726705057807.

Devine, P. (2013). The Concept of Tradition: A Problem Out of MacIntyre. *Reason Papers, 35*(1), 107–123.

Feenberg, A. (1999). *Questioning Technology*. London: Routledge.

Feenberg, A. (2009). Peter-Paul Verbeek: Review of What Things Do. *Human Studies, 32* (2): 225–28. https://doi.org/10.1007/s10746-009-9115-3.

Foucault, M. (1988). *Technologies of the Self: A Seminar with Michel Foucault*. Amherst, MA: The University of Massachusetts Press.

Gotham, K., & Staples, W. G. (1996). Narrative Analysis and the New Historical Sociology. *The Sociological Quarterly, 37*(3), 481–501.

Heidegger, M. (1977). *The Question Concerning Technology and Other Essays* (W. Lovitt, Trans.). New York, NY: Harper Perennial.

Hyvarinen, M. (2010). Revisiting the Narrative Turns. *Life Writing, 7*(1), 69–82. https://doi.org/10.1080/14484520903342957.

Ihde, D. (1990). *Technology and the Lifeworld: From Garden to Earth*. Bloomington, IN: Indiana University Press.

Ihde, D. (1998). *Expanding Hermeneutics: Visualism in Science*. Evanston, IL: Northwestern University Press.

Kaplan, D. M. (2006). Paul Ricoeur and the Philosophy of Technology. *Journal of French and Francophone Philosophy, 16*(1/2), 42–56.

Latour, B. (1993). *We Have Never Been Modern* (C. Porter, Trans.). Cambridge, MA: Harvard University Press.

MacIntyre, A. (2007). *After Virtue: A Study in Moral Theory* (3rd ed.). Notre Dame, IN: University of Notre Dame Press.

Marx, K. (1976). Theses on Feuerbach. In K. Marx & F. Engels (Eds.), *Collected Works Vol. 5: Marx and Engels 1845–1847* (pp. 3–5). New York, NY: International Publishers.

Marx, K. (1977). *Capital: A Critique of Political Economy* (Vol. 1). New York, NY: Vintage Books.

Mauss, M. (1973). Techniques of the Body. *Economy and Society, 2*(1), 70–88.

Meretoja, H. (2014). Narrative and Human Existence: Ontology, Epistemology, and Ethics. *New Literary History, 45*(1), 89–109. https://doi.org/10.1353/nlh.2014.0001.

Mitcham, C. (1994). *Thinking Through Technology: The Path between Engineering and Philosophy*. Chicago, IL: The University of Chicago Press.

Ricoeur, P. (1983). *Time and Narrative—Volume 1* (K. McLaughlin & D. Pellauer, Eds., Vol. 91). Chicago, IL: The University of Chicago. https://doi.org/10.2307/1864383.

Ricoeur, P. (1992). *Oneself as Another* (K. Blamey, Ed.). Chicago, IL: University of Chicago Press.

Romele, A. (2020). Technological Capital: Bourdieu, Postphenomenology, and Philosophy of Technology Beyond the Empirical Turn. *Philosophy & Technology*. https://doi.org/10.1007/s13347-020-00398-4.

Strawson, G. (2004). Against Narrativity. *Ratio, 17*(4), 428–452.

Taylor, C. (1971). Interpretation and the Sciences of Man. *The Review of Metaphysics, 25*(1), 3–51.

Van Den Eede, Y. (2011). In between Us: On the Transparency and Opacity of Technological Mediation. *Foundations of Science, 16*, 139–159.

Verbeek, P. P. (2005). *What Things Do*. University Park, PA: Pennsylvania State University Press.

Winner, L. (1980). Do Artifacts Have Politics? *Daedalus, 109*(1), 121–136. https://doi.org/10.2307/20024652.

Zwier, J., Blok, V., & Lemmens, P. (2016). Phenomenology and the Empirical Turn: A Phenomenological Analysis of Postphenomenology. *Philosophy & Technology, 29*, 313–333. https://doi.org/10.1007/s13347-016-0221-7.

3

Ricoeur's Hermeneutic Concept of Technical Practice

The focal point of technology ethics is *technical practice*, in its many variations. Technical practices in their multitude mediate both virtues and vices; and bring about both right and wrong conduct. At the same time, technology ethics requires an understanding of the mode of technical practice that would be the ethical aim to aspire to; a notion of a *virtuous* technical practice. These considerations require a dual but integrated understanding; of technical practice as a human activity on the one hand, and technical mediation on the other. Here, we focus on the first part of this synthesis. We place our investigation within a "hermeneutic" frame, which, in line with Romele's characterisation of "digital hermeneutics", implies that we consider meaning as something that arises out of a context of use of technologies (Romele 2019, p. 38). More broadly, we can state that an understanding of the ethical impacts of technologies follows from a more primary understanding of the way in which technologies mediate the meaning of the practices that shape our lifeworld. Accordingly, the aim of this chapter is to provide a preliminary hermeneutic concept of technical practice.

We cannot simply take for granted that a technical practice is something like a human activity involving the use of technical objects, for this

W. Reijers, M. Coeckelbergh, *Narrative and Technology Ethics*,
https://doi.org/10.1007/978-3-030-60272-7_3

will tell us nothing about the roles of narrative and of virtue with regards to this activity. Instead, as argued in Chap. 2, we need to gain understanding of how both virtue *and* narrative fundamentally relate to technical practice. For this, we turn to Paul Ricoeur, who will remain our primary guide for the remainder of this book. Ricoeur distinguishes himself from other classical hermeneuticists in the philosophy of technology, Heidegger most notably, because he paid attention to the "long route" through linguistic and material externalisations to reach an understanding of the (ethical) self (Romele 2019, p. 7). Ricoeur claims that we have no direct understanding of the self and its activities, but only an indirect, mediated one. As we will see in Chap. 4, this implies that technologies are an important aspect of the mediations of the self in our current time.

In what follows, we first provide an introduction to the work of Ricoeur and illustrate its preliminary relevance for our endeavour because of its focus on language and the social. We then turn to a problematisation of the everyday understanding of practice, which will lead us to understand the centrality of the problem of the nesting of finalities. Finally, we present Ricoeur's theory of practice as a convincing answer to this problem. This theory traces the concept from basic action, via life plans, to the narrative unity of life

Turning to Ricoeur

In Chap. 2, we argued that the work of Ricoeur would offer a promising starting point for thinking about technical practice. In this section, we expand on this argument by introducing Ricoeur's philosophical perspective and its primary position as a philosophy of mediation. We also discuss Ricoeur's treatment of language, temporality, and the social; addressing the primary lacunae in contemporary philosophy of technology.

The first thing that will strike any reader of Ricoeur's work is the immense breadth of themes, perspectives, and fields of research that it touches upon—from linguistics, via history, to phenomenology. Yet, it has a consistent focus. Throughout Ricoeur's work runs a concern with a philosophical anthropology that centres in on the critique of the Cogito as introduced by Descartes, which presumed "immediate certitude and self-transparency" (Jervolino 1990, p. xxi). The goal of Ricoeur's philosophical anthropology is to provide an alternative account of the self and simultaneously of the social world in which oneself is confronted with another (Kaplan 2003, p. 9). Opposed to the Cartesian notion of the self, Ricoeur posits a notion of the self as *indirect*, which characterises hermeneutics as a philosophy of "detours" (Ricoeur 1992, p. 17). A detour, in Ricoeur's work, consists of the problematising of a philosophical position that needs to be resolved by invoking another philosophical position that might at times seem completely foreign to it, for instance problematising Augustine's conception of time by invoking Aristotle's in *Time and Narrative* (Ricoeur 1983b). Philosophical reflection, according to Ricoeur, departs from the "fullness of language" that is always already there in human experience and understanding (Kaplan 2003, p. 20). A human does not find herself having immediate, direct access to a world, using language as a layer on top of this world to refer to it. Instead, her experience and understanding of the world are mediated through language and can therefore only be understood by means of a detour through language, involving aspects such as symbols, metaphors and narrative.

Even though Ricoeur did not construct a closed philosophical system, there is a clear coherence in his thought and the way in which he approaches his concern with philosophical anthropology. That is, he has traced the way in which language mediates one's understanding of oneself and one's lifeworld through different stages of linguistic operation. Initially, Ricoeur focused on the *symbol* as a theme of philosophical reflection (Ricoeur 1967). Because symbols operate in the public realm and are socially established and maintained, we can have no "direct" knowledge of them. A symbol is an expression that does not merely express what a speaking subject means to say but also designates something that lies beyond the speaker's intention, therefore having a *double meaning*. To demonstrate this point, Ricoeur discusses how the traditional symbolism of evil in the Western world, conceived as guilt and sin (a violation of divine command) contains a more primordial understanding of evil as defilement (an expression of chaos in nature) (Lai 1984, p. 317). Therefore, when we invoke the symbolism of evil through the notion of sin, we indirectly invoke its more primordial meaning of defilement, which nonetheless remains hidden in the common use of the symbolism. Hence, symbols in use carry with them a "surplus of meaning" that relates to their histories and often-mythical origins. They reveal meaning and at the same time conceal meaning, and the task of hermeneutics is to retrieve the hidden meaning concealed in symbols.

From the symbol, Ricoeur moved to investigating the role of *metaphor* in the creation of meaning in language (Ricoeur 2004). Metaphor can be understood as a heuristic fiction that redescribes reality (Kaplan 2003, p. 48). A famous example of a metaphoric expression is: "all the world's a stage, and all the men and women merely players" (Shakespeare 1623, p. 52). A more contemporary example would be a sentence such as "the human mind is a computer". By redescribing our idea of the world through the heuristic fiction of the theatrical stage, or our idea of the human mind through the heuristic fiction of a computer, a new possibility of *being in the world* is opened up. This possibility reveals the capacity of language to *create* meaning, to be productive. As such, a metaphor always carries with it a tension between its literal meaning ("the computer") and its imaginative meaning (the human mind understood *as* a computer). From his discussion of metaphor, Ricoeur progressed to create an account of how the rediscription of reality could transform our being-in-the-world.

For this, he turned to an even higher level of linguistic mediation, of *narrative*. Metaphor and narrative remain intimately connected because they both indicate the synthesis of heterogeneous elements. According to Kaplan, "in metaphor it is the ability to perceive resemblance in difference; in narrative it is the ability to construct a meaningful organisation of otherwise unrelated events" (Kaplan 2003, p. 50). A narrative, such as a folktale, gathers together a heterogeneous collection of events and characters, that are configured in a meaningful whole that we refer to as a "plot". Ricoeur argues that we can understand how textual narratives redescribe reality in which meaningful action takes place. By taking Ricoeur's work as a starting point for thinking about technology and technical practice, we argue in line with Lewin that "we are interested in technology first and foremost because we are interested in human freedom and agency" (Lewin 2012, p. 59). Ricoeur's ongoing project is searching for a philosophical anthropology, and any corresponding theory of technology will not be just "about" technology or technological objects, but about technology's role in the being in the world of human beings. Because we need to proceed through the long detour of symbols, metaphors and narrative to understand what and who a human being is, we need to perform the same task to understand the human as a "tool-making" being.

What makes Ricoeur's theory particularly relevant for our purposes is his general focus on *mediation*, in particular linguistic mediation of action. In relation to technological mediation, Ricoeur's philosophical anthropology can be considered to be close to the one put forward by Georg Simmel (1900). Simmel considers the human, "paradoxically, as an indirect being" (Simmel 1900, p. 227). Humans, according to Simmel, are toolmaking, technological animals, neither being determined by mechanical instincts, nor having unmediated powers and access to the world (Simmel 1900, p. 226). Ricoeur, in a similar vein, takes the notion that there is no immediate access to the self, leading to the conclusion of the non-coincidence of man with himself (Jervolino 1990, p. 15). Humans, for Ricoeur, are essentially fallible in the sense that conflict, particularly between interpretations of the self, is the most primordial (original) constitution of the self (Jervolino 1990, p. 18). Simmel similarly acknowledges this fundamental reality of conflict that arises from the antinomy between the movement of life and the stability of form (Gangas 2004, p. 18). For both Simmel and Ricoeur, therefore, humans are intermediate beings whose access to the world, to others, and to themselves, is in each case mediated. They are philosophers of mediation par excellence.

Additionally, we should acknowledge Ricoeur's philosophy not just as a philosophy *of* mediation, but also a *mediating* philosophy. That is, his work constantly mediates between different philosophical traditions and scholarly traditions outside of the boundaries of philosophy. This aspect of his work is important to consider, because it points at the potential for bringing disparate ideas together that each have their value in our discussions. As Davidson argues, "a philosophy of mediation" … "must begin by setting the existing viewpoints (the endoxa), which are all too often disengaged from one another, side by side" (Davidson 2006, p. 2). In line with this characterisation, Ricoeur consistently follows "the thought of the other to its end in order to surpass that thought" (Kemp 1996, p. 41). In his writings about language, Ricoeur incorporates ideas from both the Anglo-American tradition (commonly referred to as "analytic") and the European tradition (commonly referred to as "continental"); which motivates him to go as far as to compare the works of two philosophers who are at face value might be considered each other's opposites: Husserl and Wittgenstein (Ricoeur 2014). Analytic philosophers of language significantly influenced Ricoeur, who draws from Austin's and Searle's speech act theory and Davidson's agentless semantics of action. At the same time, Ricoeur engages with key scholars in the traditions of phenomenology and hermeneutics, such as Gadamer and Levinas. Moreover, Ricoeur draws from the work of historians such as White, semioticians such as Greimas and literary theorists such as Frye.

This broad and inclusive character of Ricoeur's work does not only build bridges between different philosophical traditions, but also establishes a rich palet of philosophical critique. While building on the works of a great variety of thinkers and drawing connections between their theories, Ricoeur remains critical at all times and incorporates no theoretical insight at face value. As such, we argue that Ricoeur is the philosopher per excellence to remedy polarising tendencies in contemporary philosophy that also affect philosophical accounts of technology, by constructing a critical philosophical mean between extremes.

In Chap. 2, we established that theories in contemporary philosophy of technology insufficiently deal with either aspects of linguistic mediation or with of collective intentionality (the social). Consequently, in order to justify the turn towards Ricoeur's work on narrative we have to show that it deals with language and the social. Unlike philosophers who predominantly focus on the material dimension of technological mediation, Ricoeur has a primary, though *not* exclusive, interest in human language. In his interpretation of works of language, he especially focuses on the "*grasp of language on experience*" (Ricoeur 2014, p. 29), on the ways in which human language mediates human experience. To illustrate this grasp of language on experience, we can consider our knowledge of traffic rules (when *such-and-such* a situation occurs, I react like *this*) as a determinant of our experience of a traffic situation, or our knowledge of ritualistic protocols during the experience of a ceremony (e.g., the installation of a new president occurs when *such-and-such* declarations are uttered). In both these cases, the way we experience the particular situation or event is mediated by our understanding and use of language.

Central to Ricoeur's understanding of language lies the interpretation of text, which he argues can be seen as a model for meaningful *action*. Ricoeur states that the object of the human sciences is meaningful action, and that meaningful action is understood by considering the discourse that shapes it, which in turn is fixed in writing—in the paradigmatic unit of a text (Ricoeur 1971, p. 322). Like textual interpretation, the interpretation of action is a dialectic of explanation and understanding because like texts, actions are readable, having a meaning that is distanced from the intentions of the actors, and subject to conflicting interpretations (Kaplan 2003, p. 68). In order to investigate how human action becomes meaningful through discourse, both spoken and written, we need to study how meaningful action is made intelligible in written texts, through language. More specifically, Ricoeur argues that humans interpret their every-day actions as configured by narratives, of which texts are the paradigmatic reifications (Meretoja 2014, p. 98). A clear example of such a text is a diary, in which personal interpretations of everyday action are recounted and thereby re-interpreted (Hassam 1990).

Ricoeur's ideas about language are strongly influenced by Heidegger, though he is also sympathetic to formalist theories that aim at providing atomistic explanations of language. A major premise that Ricoeur employs throughout his work is that human experience mediated by language is characterised by temporality: by the "within-time-ness" of human experience. Human experience and action, according to Ricoeur, are essentially mediated by language within a temporal setting, and hence are not dependent on direct access to point-like, unmediated experiences of the present. In *Narrative Time*, Ricoeur asserts: "my first working hypothesis is that narrativity and temporality are closely related—as closely as, in Wittgenstein's terms, a language game and a form of life" (Ricoeur 2014, p. 169). As such, when we consider the "time" of an action (e.g., "it's *time* to go to work"), our temporal understanding of the world is mediated by language in its narrative mode. Just how closely narrativity and temporality are related is the major subject of discussion in Ricoeur's seminal work

Time and Narrative, which will be the major source for constructing our narrative account of technological mediation (Chap. 4).

At this point, we have shown what Ricoeur potentially has to offer to address the neglect of language and temporality in contemporary philosophy of technology. But what role does the social play in Ricoeur's work? Throughout his work, Ricoeur stresses the importance of interpreting human existence by considering its embeddedness in human social reality. In his work, "the mediation of action by the imagination" as recounted in the narrative mode, "links individual with social action by bridging the realms of discourse and action" (Kaplan 2003, p. 81). In line with the "within-time-ness" of human being, Ricoeur argues: "the time of narrative is *public* time" (Ricoeur 1980, p. 175—emphasis added). Moreover, he asserts that: "public time" … "is not anonymous time of ordinary representation but the time of inter-action. In this sense, narrative time is, from the outset, time of being-with-others" (Ricoeur 1980, p. 188).[1] Thus, Ricoeur asserts that the social is explicitly present in narrative time, which remains a time of interactions between people without being made entirely anonymous, separated from human experience and action (as for example can conversely be said about the time of natural science that is derived from natural laws rather than from human experience).

Consequently, Ricoeur characterises narratives as normative, cultural phenomena by accounting for ways in which our interactions with narratives are embedded within our cultural, temporal existence. He explains why narratives can configure social reality: because they configure narrative plots that refigure social events (Borisenkova 2010, p. 93). Emplotment, which is the process that defines any narrative structure, has an outspoken normative character because the characters in a narrative are not just neutral "doers" them but are "endowed with ethical qualities" (Ricoeur 1983b, p. 59). Unlike generalised "doers" like the rational economic man who figures in economic theories, acting according to coherent, non-normative motives, characters in narratives can be

[1] Remarkably, Van den Eede's critique of Postphenomenology uses a term that is almost similar to Ricoeur's "being-with-others": the technological mediation of "being-with-each-other" (Van den Eede 2010, p. 140). This provides another rationale for turning to Ricoeur.

good or evil, rational and irrational; the protagonists or antagonists of the narrative structures. These features of Ricoeur's theory enable him to go beyond formal approaches such as the one of John Searle (2010), and to provide a holistic, normative account of linguistic mediation of the social world.

Taken together, we can convincingly draw two conclusions with regards to the use of Ricoeur's work: (1) that Ricoeur's philosophy can be characterised as both a philosophy of mediation and a mediating philosophy and (2) that it consistently pays heed to the linguistic, temporal, and social aspects of mediations of the human lifeworld. These conclusions support our use of Ricoeur's work in response to the problems of theories of technological mediation we encountered in Chap. 2, and pave the way towards our upcoming task, of connecting Ricoeur's philosophy with a notion of *technical practice.*

Problematising Technical Practice

This section problematises the everyday understanding of technical practice, and subsequently reflects on the meaning of a "virtuous" technical practice. Doing so will provide us with the final rationale for turning to Ricoeur's theory of practice, and the central notion of the "nesting of finalities" that it addresses. In starting to think about the concept of a technical practice, we reason from the everydayness of human action. This everydayness provides us with an initial, empirically informed notion of what a technical practice is. Technical practice shows itself in a great variety of ways in everyday dealings with technologies. It is most present perhaps as instances of *use* of technologies. We use a bike or a car to drive to work; we use our smartphone to navigate the roads; and we use a drill in our house to make holes for decorating the wall. In other instances, technical practices appear as instances of *design* of technologies. A software engineer codes or designs a new piece of software to update a programme, a pharmacist designs a new type of medicine to treat a decease, and a financial engineer designs a new high frequency trading algorithm to beat his competitors in the market. Finally, technical practices also appear as instances of *governance* of people or systems. A state might govern its citizens through the deployment of surveillance cameras, a manager might try to govern the behaviour of workers by giving them wearables, and an engineer might install a safety mechanism to govern the functioning of a power plant.

Hence, technical practices are extremely heterogeneous. How, then, can we understand them in a more general way? Philosophy of technology already brings us part of the way in giving an answer. Drawing from Heidegger, Don Ihde established that all technical practices involve a "use context" (Ihde 1979, p. 119), which means that they are embedded in a totality of useful things and relations; what Heidegger calls a "totality of relevance" (Heidegger 1996, p. 80). This use context is structured according to the "in-order-to" relation (Heidegger 1996, p. 78), which gives it a certain variance that Ihde calls "multistability" (Ihde 1990, p. 144). What this means is that I might use a technology like a smartphone in different ways: "in order to" find the way on Google maps, "in-order-to" call a colleague, or "in order to" send an email to a friend. Technologies enable a very heterogeneous set of activities, but still have the relation of the "in-order-to" in common.

For Heidegger, the "in-order-to" is only one of the three structural elements of what is called "worldliness" (Heidegger 1996, p. 61), which is constitutive of our being-in-the-world. The other two elements are the "where-from" (the materials from which tools and technologies are produced), and the "for-the-sake-of-which" (the end of a technology in use). These latter elements have unfortunately been largely neglected in contemporary philosophy of technology (Reijers 2019). To understand the whole structure of a technical practice, we return to the place where Heidegger initially took his notion from, namely Aristotle's *Nicomachean Ethics*. There, technical practice is determined out of the concept of *technê*, which designates the virtue of *how* things need to be done, a "knowing-how". What Aristotle adds to our understanding is that technical practice is not just an action in order to achieve something else, but also that it has a sense of completion, or *telos*. Hence, a practice has a certain state of "being-completed"; consider the state of health when the doctor has treated a patient. It also means, simultaneously, that every practice has a *limit*, as that beyond which it knows no further end. For instance, the practice of healing knows no further end beyond the achievement of health of the patient.

However, here we hit upon a first problem with the concept of a technical practice. Consider for instance a shoemaker, who fabricates a shoe. This shoe has a *telos* of being worn, but this telos is not at all included in the practice of shoe making. That is, the hitting of the nails, the sewing of the leather, and so forth, all seem unrelated to the activity of wearing the actual shoe. This is where we encounter the phenomenon of a *nesting of activities*. For instance, I might use a hammer in order to hit a nail in the wall. This, I might want in order to fasten something, which in turn is done in order to secure the house against bad weather. Finally, this is all done for the sake of providing shelter (Heidegger 1996, p. 78). Eventually, through the context of use of many different technologies, we refer back to a possibility of being, a way for us to be in the world. Crucially, this concerns a possibility of being as living together in a political community; of social life. A technical practice is therefore always a social activity, even when it is engaged with in a solitary manner.

At this point, we have established an initial structure of technical practice that allows us to understand this "mode of being" as a whole, out of a heterogeneous set of different activities. Differently put, what technical practices have in common is that they all are placed in a totality of relevance, or use context, in which they stand in relation to other technical practices in according to the "in-order-to" structure and refer to a "for-the-sake-of-which", which is a telos of the practice with regard to a mode of being of "Dasein", a being-with-one-another in a political community.

This, however, brings us to a second difficulty. Namely, the structure of a technical practice presented above does not tell us anything about the distinction between a technical practice that is either conducive for or obstructive to the cultivation of virtue. More plainly said, we have no way—yet—to distinguish between "good" and "bad" practices. If there is something like a virtuous technical practice, we need to be able to distinguish it from its opposite. Returning to MacIntyre, we recall the distinction between internal and external goods. For instance, a concert pianist can engage in her practice of piano playing by aiming at its internal good of *mastery*, thereby cultivating her virtue as a musician. Yet, the pianist does not *need* to do so, and can also give a sloppy, bad performance. In our common understanding, we would say that in order for the pianist to play in virtuous manner, she would need to be *awake* and *succeeding* (Heidegger 2009, p. 60). That is, she needs to be focussed on her activity continuously and succeed in every way, not playing any false notes or miss the rhythm; otherwise the activity would not be praiseworthy as virtuous. As such, the meaning of virtue can likewise be captured by the notion of "seriousness" (Heidegger 2009, p. 121).

A virtuous technical practice must thereby be distinguished from the mere application of a technique or "training", which allows for errors and "sleepiness". To illustrate this difference, consider the example the technical practice of driving a car. When learning how to drive a car through training, which includes techniques such as the correct shifting of the gears, one is expected to make errors initially and to learn from those errors towards perfecting the technique. The instructor is ultimately responsible and guards against sleepiness or errors of the student. However, when practising driving on the road in the environing traffic (not in the sense of training, but of actual doing), one is not merely expected to apply the proper technique but to be awake, in this sense of having a determinate possibility to be resolved to act in a certain way in a definite moment (for instance, hitting the brake when a pedestrian suddenly crosses the street, while simultaneously being aware of the car approaching at the back), and to succeed in each case, meaning for instance that hitting the pedestrian as an "error" is in no case praiseworthy—even if one "learns" from it.

Thus, "good" and "bad" technical practices, or rather virtuous and non-virtuous technical practices can be distinguished through consideration for the disposition of the agent. Non-virtuous practices allow for sleepiness and are characterised by the possibility of success and error. These practices are not "bad" in any categorical sense, and actually important as a part of learning or training. For instance, a computer science student who is learning how to code can learn a great deal from making mistakes once in a while, accidentally coding "bugs" that make programs dysfunctional. In contrast, virtuous technical practices require a disposition of seriousness, of being awake and succeeding. Simply said, seriousness is required when the "stakes" are high. This is characterised by the involvement of a practice in a totality of relevance that has a particular "for-the-sake-of-which" in social life. Only with this type of technical practices comes a distinct sense of responsibility, and therefore a genuine possibility of both virtue and vice. Consider for instance a technical manager responsible for delivering the code of a new payment system that will be used by millions of people. The technical practices of governing this process require her to be awake and not to allow for error; for this will in no case be praiseworthy.

We here arrive at a third difficulty of conceptualising technical practice. That is, even though a technical practice can be understood according to the idea of a totality of relevance, and a "good" technical practice according to the notion of seriousness, we hit upon an enigmatic indeterminacy. Namely, the notion of "good" in a technical practice can easily be carried over from a "good" human being to for instance a "good" assassin. An assassin could be very serious in carrying out his technical practices (e.g., assassinating a politician with a sniper gun from a far distance), and also link the in-order-to of these practices to a for-the-sake-of-which of his political community (e.g., a clan of assassins). Yet, we do not want to call this type of practice "virtuous", since the killing of other people for money cannot be anything else than a vice. In other words, technical practices need something more than simply to be awake and succeeding in order to be virtuous in a genuine sense.

This third difficulty cannot be fully addressed at this point and will require the remainder of this book to be fully unpacked. What we will do at this point, in order to find a *way* to tackle this problem, is to show that (1) this problem is captured by the notion of authenticity in the "nesting of finalities", (2) authenticity requires and understanding of temporality, and (3) this understanding can be gained through narrative theory. For this, we consider Ricoeur's confrontation with Heidegger's phenomenology of Being.

First, Ricoeur follows Heidegger in considering the nesting of the in-order-to with regards to the for-the-sake-of in technical practices, in the sixth study of *Oneself as Another* (Ricoeur 1992). He states that in his account of practice, which explicates the "relation between practices and life plans", the "secret of the *nesting* of finalities, one inside the other", can be found. He argues that this responds to a central difficulty in the *Nicomachean Ethics*, resulting from the paradox that "each praxis has an end in itself and that all action tends toward an ultimate end" (Ricoeur 1992, p. 178—emphasis added). The ultimate end, for Aristotle, is *eudaimonia*, which does not only denote a final end of a practice in relation to social life, but also a state of being "whole" (i.e., being-completed as the being that we ourselves are). In other words, virtuous technical practice needs to refer through a totality of relevance to the end as being-a-whole of the beings that we ourselves are. This mode of being is called *authentic*. Originating from Heidegger, authenticity denotes the German word "eigentlich", which can perhaps better be understood as "ownedness"—denotes a mode of being which has a telos that belongs to Dasein itself.

Second, Ricoeur continues to follow Heidegger in considering that the key to authenticity lies in an understanding of *temporality*. The primordial experience of temporality that is at stake in virtuous being, according to Heidegger, derives from the "being-a-whole" (i.e., being-completed) of Dasein, which in turn is experienced in anticipatory resoluteness (Heidegger 1996, p. 281). Anticipatory resoluteness, as an authentic mode of being, is ultimately linked to the temporal "being-towards-death" (Heidegger 1996, p. 353), in which he sees the outmost possibility of a human life (the possibility of life's impossibility) and therefore the ultimate telos or being-completed of Dasein. Telos, in this understanding, is "being-toward the end in such a way that this end constitutes the genuine there, determining, in a genuine [authentic] way, a being in its presence" (Heidegger 2009, p. 92). Heidegger thereby completes the circle: virtuous technical practice, as being awake and succeeding, is an authentic mode of being insofar as it is in the mode of anticipatory resoluteness in the face of being-towards-death, the being-completed—or telos—of Dasein.

Yet, like Ricoeur, we will not follow Heidegger any further here. Contemporary philosophy of technology has provided persuasive reasons for this decision. As Verbeek for instance argues, contra Heidegger, "a philosophical analysis of the role of technology in the modern world cannot rest with reducing technology to forms of interpretation, but needs to devote its attention as well to the ways in which specific technologies and artefacts help to shape specific forms of praxis and interpretation" (Verbeek 2005, p. 67). The motivation for Verbeek's claim can be derived from Heidegger's way of conceptualising authenticity. To address the problem of "being-a-whole", Heidegger allows us only to retrieve the authentic through the individualistic notion of being-towards-death. This notion is individualistic because Dasein "is completely thrown back upon its ownmost potentiality-of-being" (Heidegger 1996—emphasis added) in being confronted with being-towards-death and is thereby so-to-say completely on its own in coming to terms with this possibility of death. The inauthentic, in contrast, is constituted in "average everydayness" (Heidegger 1996, p. 235): in the flight from death instead of in its resolute anticipation, which reveals itself in the public realm, in which "the They"—the anonymous other—is encountered. Heidegger thereby juxtaposes the everydayness of public life, politics and technology in which Dasein's being-in-the-world has the character of "falling prey" to something (Heidegger 1996, p. 164), with the authentic being-in-the-world of Dasein, which has to assume the mode of anticipatory resoluteness in the face of being-towards-death in order to attain authenticity. In contrast with Heidegger's ideal of authenticity, practising philosophy and ethics of technology inescapably takes place in everydayness, in the public realm of research labs, teams of engineers working on research and innovation projects, and political interference in innovation processes. Approaching the complexity of this everydayness from the solitary starting point of being-towards-death seems a highly impractical endeavour.

However, even though this critique provides us with an initial motivation to search for alternatives, it does not yet offer sound philosophical reasons for denouncing Heidegger's notion of authenticity. We also cannot simply ignore the problem of authenticity, as most of contemporary philosophy of technology has done. The reason for this is that authenticity, as we saw, denotes the possibility to distinguish between a genuine virtuous technical practice and other modes of this human activity. Ricoeur, in contrast to philosophers of technology, did pay attention to the problem of authenticity, so much so that he appears to have built his entire theory of practice around it.

Ricoeur understands that the structural wholeness that is at stake in "being-a-whole" (the *telos* of Dasein) led Heidegger to the question of authenticity and thereby of temporality. His formidable critique of Heidegger's philosophy departs from an original proposition: namely from stating that the unfinished account of phenomenological time in

Being and Time is an outermost attempt to solve the aporias of time (Ricoeur 1988, p. 92). In other words, because Heidegger ultimately hits a limit of a complete account of temporality, he cannot provide a full account of authentic virtuous practice, either. Ricoeur first of all recognises the crucial point that for Heidegger's project to succeed: the "conquest of primordial concepts is" … "inseparable from a struggle against inauthenticity", against everydayness (Ricoeur 1988, p. 65). However, Heidegger's existential analysis has to be based on a testimony of everyday existence, which leads to a lack of a distinction between these two modes of understanding (authentic/inauthentic) if the grounding of authenticity does not succeed. Consequently, Ricoeur questions Heidegger's conception of authenticity as deriving from being-towards-death and argues that competing tests of authenticity (for instance, authenticity as being grounded in "being-born" in the work of Arendt) are equally legitimate (Ricoeur 1988, p. 67). He even considers an analysis such as Sartre's as an equally legitimate limit case, which considers death an interruption of the human potentiality-for-being instead of its being-completed.

Ricoeur further compromises Heidegger's notion of temporality on which his understanding of authenticity depends, by criticising Heidegger's unavoidable polemic against the ordinary, everyday concept of time that alone enables a conception of the opposite, authentic concept of temporality. He discusses how Heidegger conceives of the ordinary concept of time as a result of the "levelling off" of the authentic understanding of temporality through datability (situating an event in the indifferent "now"), lapse of time (the interval between "since then" and "until") and the publicness of time of preoccupation (the concealedness of our authentic experience of time behind an average, public understanding of time as a series of point-like "nows") (Ricoeur 1988, p. 84). Heidegger argues that by publicly "dating" moments in time, using clocks and calendars, and by understanding time as an interval between dates (e.g. "the meeting took place between 1pm and 2pm"), we cover up our original, phenomenological understanding of time. Ricoeur follows Heidegger, in accepting that a derivation of human temporality from a series of point- like "nows" (as science might claim) is impossible. Yet, he confronts Heidegger by stating that the reverse—the derivation of cosmic time (the time of the world) from a phenomenological account of temporality and Dasein is "just as impracticable" (Ricoeur 1988, p. 88).

Contra Heidegger, Ricoeur recognises Aristotle's conception of cosmic time, as having something to do with movement in the world, as being equally valid as Heidegger's phenomenological notion of temporality, derived from being-towards-death. In fact, Ricoeur's entire endeavour in *Time and Narrative* seems to be a response to the aporias of time that were opened up by his critique of Heidegger. He states that "having something to do with movement and something to do with Care [praxis] seem to me to constitute two irreconcilable determinations [of time] in principle" (Ricoeur 1988, p. 89).

Third, Ricoeur leads us to *narrative* as the likely solution to the aporias of temporality that need to be addressed in considering technical practice. In denouncing Heidegger's perspective, Ricoeur rehabilitates modern scientific contributions to our understanding of time, those coming from geology, evolutionary theory and astronomy, which contribute a legitimate "diversification in the meanings attached to the term 'time'" (Ricoeur 1988, p. 90), meanings which are not reducible to an all-encompassing notion of time as a series of "nows". He states that our ordinary conception of time encompasses both existential consideration (i.e. "time flies" because we flee from death) and cosmic considerations (i.e. "life is short" because of the immensity of cosmic time). These two ways of grasping time overlap in our ordinary, everyday conception of time, which in turn makes us attentive to the incommensurable polarities they entail; generating paradoxes of time. But if this is the case, how then do we make sense of time altogether? Ricoeur puts narrative forward as the answer. He argues that the plot in narrative can be considered as a narrative matrix (Ricoeur 1979, p. 23), which contains temporal structures; both chronological (one after the other) and non-chronological. First, narration extends through time in our everyday existence through "existential traits" like notions of "then" and "later". Second, narratives situate actions "in time", linking events together, creating a sense of ending and generating a memory of reading. Third, narrative as history enables us to engage in repetition of action, of linking our heritage to our practices. This connection between narrative and time will be further explored in Chap. 4. What is important to consider here, is that Ricoeur puts narrative forward as an alternative to Heidegger's existential notion of being-towards-death; and therefore, as a promising starting point for addressing the *secret* of the nesting of finalities in technical practices.

Where did this all bring us? The problematisation of the everyday understanding of technical practice has given us an initial understanding of the concept of a *virtuous* technical practice. That is, a technical practice needs a sense of being-completed to be considered virtuous, which is captured by the notion of authenticity. Authenticity makes the temporality of a technical practice that cultivates the virtues explicit, which distinguishes it from merely applying a technique, and calls for an understanding of narrative as that which responses to the polarities in our understanding of time. More simply said, we could state that whenever we engage in a technical practice virtuously, we need to be aware of how this practice contributes to the bringing to completion of human life in the social world, which is done through the narrative mode. For instance, a saddle maker in Ancient Greece needed to be aware (a knowing *why*, not merely a knowing *how*) of how his saddle making contributed to the war planning and life in the polis, and be awake and succeeding in his practice by resolutely anticipating the type of war that will be fought, which might require him to produce lighter or heavier saddles.[2] We thereby return to MacIntyre's use of narrative, namely as that by which we explain our actions. However, Ricoeur positions narrative more firmly within his theory of practice, by inserting narrative theory at two levels of analysis: at the level of practice, where it makes basic actions and action-chains intelligible *and* at the level of life plans, where it makes practices intelligible with recourse to ideals. In what follows, we turn towards Ricoeur's theory of practice in *Oneself as Another* (1992) that we will reinterpret as a theory of technical practice.

A Hermeneutic Account of Technical Practice

Ricoeur's theory of practice introduces a number of core concepts: of practices, life plans, and the narrative unity of life. We will reflect on each of these concepts, how they relate to the narrative mode, and how they together form an integral conceptualisation of technical practice. To

---

[2] Even though we use an example of an ancient craft, this illustration brings us very close already by what we might mean with "responsible research and innovation".

understand the premises that underlie Ricoeur's theory of practice, we will first turn to Arendt's theory of action (1958).

Reflecting on Arendt's theory of action, which encompasses the human activities of labour, work and (ethical) action, Ricoeur accredits her with two important discoveries (Ricoeur 1983a). First, Arendt distinguishes between the central human activities of labour, work and action by considering their *temporality* (Ricoeur 1983a, p. 61). That is, labour is a timeless activity, in that it is concerned with nutrition (consumption) and the "ceaseless renewal of life" (Ricoeur 1983a, p. 63). Work or fabrication, in contrast, is the source of durability in time—of things that continue to exist beyond the perishing of their makers. Praxis, finally, as an activity initiated by a human actor, is not durable but essentially only exists as long as the actor sustains it, after which it disperses.[3] Arendt's three basic human activities therefore denote distinct temporal modes of being. Second, Arendt acknowledges the significance of narrative, by stating that praxis only reveals itself to the storyteller (Ricoeur 1983a, p. 68). In other words, it is only in the narrative mode that praxis and human temporality become intelligible, because it places the singularity of a human action as a fragile initiative within the "interplay of circumstances induced by the web of human relationships" (Ricoeur 1983a, p. 67). It seems that Arendt's theory of praxis has had a profound effect on Ricoeur's conceptualisation of practice, by stressing the factors of temporality and narrative. However, Ricoeur moves beyond Arendt by offering a different notion of authenticity (one that allows for a multitude of phenomenological interpretations of the wholeness of being) and therefore arriving at a different theory of practice.

Another significant source of inspiration for Ricoeur has been MacIntyre's theory encompassing concepts of practice, life plan, and moral community. Similar to Heidegger, MacIntyre identifies the good with telos[4] (MacIntyre 2007, p. 52). Knight, in examining the relation

[3] For this reason, we do not "make" or produce history as a work, according to Arendt, but we *act* or practice it.

[4] Knight claims the exact opposite, namely that by identifying the good with telos, MacIntyre was refuting Heidegger's philosophy (Knight 2008, p. 40). However, as we saw before, Heidegger explicitly states that a good is always a limit of praxis, and as such should be considered telos, as a

between Heidegger and MacIntyre, claims that according to the latter "following moral rules and cultivating such excellences of character as courage, truthfulness, temperance and justice is what progresses us from our 'untutored' state to the human end or telos of rational self-fulfilment" (Knight 2008, p. 40). Contra Arendt, MacIntyre draws production and action closer together, even to the point of considering the virtues as analogous to *technai*, productive skills, that are acquired in a similar way as an apprentice acquires a craft such as painting (Knight 2008, p. 41). For MacIntyre, our understanding of practice does not merely depend on some biological account of human nature, but on an account of humans as social beings that engage in *shared* practices. He therefore argues for the need of a sociological account of practices that defines them through socially established standards of excellence. Because we have these shared standards of excellence for both the outcomes of a technical practice and the activity itself, we can discuss two distinct types of goods that are internal to the practice. For instance, we have standards of excellence for painting as an activity ("she is painting very well") and for the outcome ("that is an excellent painting"), which constitute goods internal to the practice of painting. The goods internal to practices, in turn, are nested with regards to goods internal to a certain type of life, for example of "being a painter" (Knight 2008, p. 41), which constitutes a life plan. Life plans, then, are nested within what MacIntyre calls a moral community.

Even though Ricoeur sees a "felicitous encounter" (Ricoeur 1992, p. 158) between his and MacIntyre's theories of practice, he is not uncritical towards the latter. On the contrary, and in line with the criticism we developed in Chap. 2, he states that MacIntyre's notion of narrative restricts itself to the stories told in the thick of everyday life without engaging the role of fiction, of imaginative variations, and thereby of *innovation* as opposed to tradition. Indeed, it seems that MacIntyre assigns to tradition a similar role as Heidegger assigns to authenticity. A living tradition is conceptualised as a "historically extended, socially embodied argument" (MacIntyre 2007, p. 222), that provides the for-the-sake-of-which of people's practices and life plans in a moral

being-completed. However, the difference between MacIntyre and Heidegger lies in the fact that the first accepted the idea of an end, and therefore a good in itself.

community. Since the tradition is thereby the source of standards of excellence that determine the goods internal to practices, it is the differentiating factor between those practices that are inauthentic and those that are authentic. Indeed, it seems difficult to envision in MacIntyre how individuals through imaginative variations—through fiction—might challenge historically embedded standards of excellence, and thereby challenge their moral traditions as well. By means of his narrative theory, Ricoeur manages to surpass this difficulty.

To formulate a theory of practice, Ricoeur starts by positing that "narrative theory occupies a central position [...] between the theory of action and ethical theory" (Ricoeur 1992, p. 152).[5] In other words, narrative allows us to move from a description of an action (e.g., agent *x* declared *y* in order to *z*) to a prescription of action (e.g., agent *x* should not have declared *y* because of *z*). To revise the concept of praxis, he argues that "a hierarchy of units of praxis must be made to appear" (Ricoeur 1992, p. 153), each containing its own principle of organisation. This hierarchy should not only provide a series of elements or parts that together constitute the notion of practice, but should also make apparent *how* narrative theory connects the elements with one-another, making one element intelligible in terms of another. For instance, it is in the narrative mode that practice can be made intelligible in terms of a life plan: "I'm *farming* [practice] the land because it has been the calling of my family for generations and I want to sustain this *type of life* [life plan] for the next generation". Thus, the narrative mode allows the agent to explain why he engages in a certain practice in terms of his life plan. The three elements that together constitute the hierarchy of praxis that Ricoeur presents are (1) practices, (2) life plans and (3) narrative unity of life. Ricoeur dis-

[5] As an example, consider the following narrative: "Elizabeth, the nurse, pulls the plug, gathers the family members, looks at the mother and says: "may you find peace". The machines die out and a life passes by and slips away." Disjointed, these simple actions of the form "pulling x", "calling y", "saying z" and their related chains of actions do not add up to an intelligible account of the meaning of the entire plot. Only when considered in its narrative unity, the discordant, heterogeneous elements of "pulling a plug, calling the nurse" as a chronological sequence, "a life passing by" in terms of recollected memories and a life "slipping away" in terms of death as the end of time, form a synthesis. It is therefore the connection between a basic action of "pulling the plug", its relation to the practice of patient care and its ethical significance in relation to a life plan (or in this case, the end of a life plan).

cusses these elements as stages that allow us to move from a description of action to an evaluation and prescription of action. It might therefore actually be better to characterise each stage as a transition concept: from action to practice, from practice to life plan, and from life plan to the narrative unity of life, and—as we shall see—to the ethical aim.

Action < > Practice

Ricoeur puts great emphasis on epistemological concerns and does not merely ask the ontological question "what is practice?" but also the epistemological question, "how do we understand practice?" It is actually the latter question that is his point of departure, and which makes Bobb designate Ricoeur's philosophy as offering a "reversed ontology", characterised by starting from an investigation of the way we understand ourselves and the world through texts (and in our account, through interaction with technologies) and moving from that investigation ever closer towards an ontology of the self (Bobb 2011, p. 344). As a first step in his theory of practice, Ricoeur deals with the transition between the recounting of basic actions and action-chains to the narrating of the "global action" that we designate as a practice. Practice, for Ricoeur, is a rich, unifying concept that—in accord with Heidegger's notion of virtue—has to designate an activity that stretches across time. Simply playing a note or a melody is not a practice, but "playing the piano" as in performing a concerto can be considered a practice. It is therefore only with regards to the latter that we can designate a practice as "virtuous", as being-awake and succeeding. In recounting a practice, a text could for instance state: "the farmer's passion for farming became apparent in the resolute but delicate way he placed the plough in the soil to prepare the land". In this example, the verb "place" designates a basic action, the verb "prepare" an action-chain and the verb "farming" a practice. We will shortly reflect on these three concepts.

A *basic action* designates any action that can be performed without depending on another action. Such a basic action can be a gesture (e.g., waving a hand), a posture (e.g., sitting down), or more basically an elementary corporeal action. Such basic actions enable the notion of the in-order-to, and it becomes clear at this point how Heidegger's primary example of hammering involves such a structure: hammering, as a basic action of the hand, mediated by the hammer, is performed in-order-to fasten something. The in-order-to relation enables the coordination of basic actions in long *action-chains*, which link actions together in causal series in which the agent performing them can intentionally intervene (Ricoeur 1992, p. 153). This might be expressed in describing a game of chess, by explicating a strategy in terms of a sequence of basic position-changes of pieces on the board. Ricoeur additionally refers to speech act theory to explain how each basic action receives meaning through the notion of a *constitutive rule* (Ricoeur 1992, p. 154), which changes a gesture of moving a wooden object across a wooden plane into a move in the game of chess. As such, a constitutive rule results from a declaration in the form "resituating this wooden object from this space to that space *counts as* a move of the queen in the game of chess" (cf. Searle 2010). Ricoeur connects this concept of constitutive rule with MacIntyre's notion of *standard of excellence*, which opens up the possibility for evaluating an action, though not yet in a strictly ethical sense. For instance, certain constitutive rules allow one to evaluate a chain of basic actions made on a chessboard as either valid or invalid moves in the game of chess.

These rules introduce the *interactive* character of practices, because they only exist in virtue of being learned from others or evaluated by others, not merely by an isolated individual (Ricoeur 1992, p. 156). To be sure, to interact can also mean to abstain from action (e.g., refraining to help someone), which ensures Ricoeur's theory to encompass both *acting* and *suffering* beings.[6] What we can conclude at this point is that practices should neither be understood merely in terms of linear action-chains, nor according to their nesting-relations determined by constitutive rules, but always in terms of both.[7]

[6] Through being thus conceptualised, practices link up with the ethical aim in two ways: (1) by providing a basis to consider goods internal to a technical practice, and (2) by providing the "content for the empty form of the categorical imperative" (Ricoeur 1992, p. 177).

[7] In line with this finding, Ricoeur makes explicit that his analysis accords with our earlier discussion of Heidegger's notion of technical practice, by stating that through the integration of linear and subordinating (i.e. nesting) viewpoints of basic actions, action-chains and practices, "no action is only poeisis [linear] or only praxis [nesting]" (Ricoeur 1992, p. 176).

Practice < > Life Plan

The abovementioned transition from basic action to practice did not yet provide us with a complete account of practice. On the contrary, Ricoeur stresses that to understand practice properly we have to traverse the stages of life plan and the ethical aim. Ricoeur spends relatively little time discussing the notion of life plan, but this is because it introduces the movement of narrative configuration (see Chap. 4) into the fabric of practice, which he already extensively dealt with in *Time and Narrative* (Ricoeur 1983b). A life plan is not to be understood as an "entity" with definite properties such as "*a* profession", but as a movement that links practices with the wholeness of a human life. Ricoeur's choice of wording seems slightly unfortunate when he designates life plans as "*vast* practical *units* that make up professional life, family life, leisure time, and so forth" (Ricoeur 1992, p. 157). In fact, a life plan is not to be considered as a unit, as an entity, but as a particular movement of "emplotment" (the synthesis of characters and events in a plot), which is not necessarily to be considered "vast". This is the case because a life plan seems to be any movement of emplotment that links practices with the narrative unity of life, which can be either "tiny" (for instance, a narrative about visiting the zoo with one's grandfather on a particular afternoon) or "vast" (for instance, a narrative about being a zoo keeper). Both of these constitute a "life plan" for they configure plots with a beginning, middle and end that additionally are connected with one-another (for instance, one became a zookeeper because of that vital afternoon visit) without the one being subsumed under the other.

Notwithstanding this slight diversion, we retain the crucial addition provided by Ricoeur to the notion of narrative configuration in constructing his theory of practice. This addition consists in characterising the life plan as a "back-and-forth movement between more or less distant ideals" (Ricoeur 1992, p. 157). In other words, we could state that we understand a technical practice that we engage in, such as flying a miniature airplane, in terms of more or less distant ideals, such as wanting to become a pilot, which are *mediated* by life plans such as partaking in a miniature airplane competition. In this example, the airplane competition constitutes the plot that connects the practice and its standards of excellence with the distant ideal of being a pilot. It is also at this point that narrative configuration through history or fiction enters the stage, for just as recounting a past experience can connect practices with more or less distant ideals, so can works of historical narrative or literary fiction. For instance, the practice of securing one's ICT infrastructure can connect with a distant ideal of a free and open society mediated by the reading of Orwell's *1984*. The difference between Ricoeur and MacIntyre in this case is that MacIntyre only foresees a "bottom-up" movement, from practices, towards life plans and the narrative unity of a human life while Ricoeur characterises it in accordance with a "twofold principle of determination" (Ricoeur 1992, p. 158), which consists of the twofold movements of *ascending complexification* and *descending specification*. Ascending complexification denotes the movement from basic actions and practices towards ideals, for instance starting to play the piano at an early age, by pressing some keys, and moving towards the ideal of becoming a professional piano player. Descending specification denotes the movement from ideals towards practices and basic actions, as in the above-mentioned example of reading *1984*, which enables the practitioner to start from ideals and move towards a change in practice and its related basic actions.

Life Plan < > Narrative Unity of Life

After having considered life plans, Ricoeur turns to what MacIntyre designates as the "narrative unity of life" (Ricoeur 1992, p. 160): the overall life biography that supports the notion of a "good life" and provides life with its ethical tenor. As such, a narrative unity of life fulfils the for-the-sake-of-which of Dasein, which returns the telos of the in-order-to of practices back to being-with-one-another in the polis. This narrative unity is necessary for virtue, for being awake and succeeding in a practice, which requires the for-the-sake-of-which. However, Ricoeur objects to the idea that one could speak of something like a coherent and complete narrative unity of life. First, we are never the sole authors of a life, but only ever co-authors, just as we are only co-authors in our interactions with technologies. Second, unlike a narrative, a human life cannot be easily considered as "complete", for its beginning in birth and ending in death are inaccessible for a human to "gather together" in a story. Third, a life can consist of several stories, each with their own sense of ending, without necessarily being unified. Fourth, whereas different works of narrative commonly present incommensurable plots, our life stories are entangled with the plots of other people's life stories. Fifth, our recounting of life only applies to past experiences whereas works of narrative can include narrated accounts of future states of affairs. These arguments lead Ricoeur to claim that the narrative unity of life must *not* be understood as a coherent and complete whole (*as* a completed narrative), but rather as an "unstable mixture of fabulation and actual experience" (Ricoeur 1992, p. 162) that can move *towards* a sense of completion without ever attaining it. To be sure, narrative mediates the sense of unity of a human life but does not offer a complete account of it.

This conclusion necessitates a move from narrative theory to ethics, because we again encounter the problem that we earlier encountered—of being unable to distinguish the good person from the "good" assassin. Consider the practice of stealing something, mediated by life plans such as partaking in one's first armed robbery in a criminal gang, which connects it to the distant ideal of being the boss of a crime syndicate. The in-order-to of stealing something is nested with regards to the for-the-sake-of-which of being the political leader of a group. One can therefore be awake and succeeding in practising stealing, and therefore be virtuous in a certain sense. However, the problematic nature of the activity of "stealing" as a virtuous activity makes that we want to move beyond this notion of virtue, and to be able to distinguish between technical practices that cultivate virtues leading towards *eudaimonia*, and those that do not. Indeed, being a successful thief does not seem to accord with leading a good life. Ricoeur therefore still needs an equivalent for Heidegger's notion of authenticity that he finds in "the ethical aim" (Ricoeur 1992, p. 169), which we will discuss in Chap. 5.

For now, we can conclude by mentioning how Ricoeur sees that narrative leads to the ethical aim—or perhaps better the "being-completed" of ethical life. First of all, the experiences recounted in the narrative mode always involve estimations and evaluations that enable actions to be "subject to approval or disapproval and agents to praise or blame" (Ricoeur

1992, p. 164). In fiction in particular, moral judgement is thereby subjected to imaginative variations that allow practitioners to question their practices and related standards of excellence. At this point, Ricoeur connects his theory of practice with the main theme of his book, being the relation between *sameness*, which can be found in lasting dispositions of character, and *selfhood*, which can be found in keeping one's word despite change; which together constitute personal identity. Even though it is outside the scope of our current endeavour to delve into the breath of this thematic, it needs to be pointed out that the narrative mode mediates between sameness and selfhood, by narrativising the ethical aim of the good life and thereby "giving recognisable features to characters loved and respected" (Ricoeur 1992, p. 166).

Conclusion

Even though our concept of technical practice is still underdeveloped, we gained a great deal of insight through our reflections. First, we established that technical practices have a common structure, which includes a "where-from", an "in-order-to" and a "for-the-sake-of" as constitutive elements. Second, we learned that this structure confronts us with the enigma of the "nesting of finalities", and the difficulty of distinguishing between virtuous and non-virtuous practices. Third, it became clear that this difficulty needs to be addressed through an understanding of temporality, which Ricoeur argues is to be accessed through narrative. Finally, these considerations together led Ricoeur to formulate a theory of practice that consists of three stages: from basic actions to practices, from practices to life plans, and from life plans to the narrative unity of life.

However, two aspects are still crucially missing in our account of technical practice. First, the critical reader might remark that our account of practice has thus far nothing "technical" to it, in that it covers all practices. This is certainly the case, and thus far our theory of practice only deserves the predicate "technical" by virtue of having to do with the mode of being of *technê*, a "knowing-how". We have not yet discussed the role of technologies in the coming to be of technical practices, and thus not yet provided an account of technological mediation. From philosophy of technology, we have gained that technological "things" mediate our practices, and as such intervene in our practices, life plans, and movement towards a narrative unity of life. As shall become clear, technological mediation can be treated in way analogous to *narrative mediation* as present in Ricoeur's theory of practice. As such, technologies are involved, like narratives, in the movement of *ascending complexification* and *descending simplification*. Like narratives that permeate our everyday lives, technologies instantiate a "bottom-up" movement, inviting and inhibiting our practices; as well as a "top-down" movement, presenting us with distant ideals with which we reflect on our practices. Chapter 4 will be devoted to this missing aspect, by linking our theory of technical practice with an account of technological mediation that is inspired by Ricoeur's narrative theory.

Second, the final difficulty that we encountered in our problematisation of practice has still not been resolved, far from it. That is, thus far we only know that Ricoeur's notion of the ethical aim offers us a way to distinguish between virtuous and non-virtuous technical practices. However, we also got to know that this notion of the ethical aim is open ended, so to what extend does it provide us with an understanding of the "end" of Dasein as is required by the concept of authenticity? This question will be pursued in Chap. 5, in which we will further explore Ricoeur's concept of the ethical aim.

References

Arendt, H. (1958). *The Human Condition*. Chicago: University of Chicago Press. https://doi.org/10.2307/2089589.

Bobb, C. (2011). Paul Ricoeur's Hermeneutics Between Epistemology and Ontology. *Logos & Episteme, 2*(3), 335–345.

Borisenkova, A. (2010). Narrative Refiguration of Social Events: Paul Ricoeur's Contribution to Rethinking the Social. *Études Ricoeuriennes / Ricoeur Studies, 1*(1), 87–98. https://doi.org/10.5195/errs.2010.37.

Davidson, S. (2006). Thinking After Ricoeur. *Journal of French and Francophone Philosophy, 16*(1/2), 1–8. https://doi.org/10.5195/jffp.2006.179.

Gangas, S. (2004). Axiological and Normative Dimensions in Georg Simmel's Philosophy and Sociology: A Dialectical Interpretation. *History of the Human Sciences, 17*(4), 17–44. https://doi.org/10.1177/0952695104048071.

Hassam, A. (1990). "As I Write": Narrative Occasions and the Quest for Self-Presence in the Travel Diary. *ARIEL: A Review of International English Literature, 21*(4), 33–47.

Heidegger, M. (1996). *Being and Time* (J. Stambaugh, Ed.). New York: State University of New York Press. https://doi.org/10.1017/CBO9781107415324.004.

Heidegger, M. (2009). *Basic Concepts of Aristotelian Philosophy*. Bloomington: Indiana University Press.

Ihde, D. (1979). *Technics and Praxis* (Vol. 33). Dordrecht: D. Reidel Publishing Company.

Ihde, D. (1990). *Technology and the Lifeworld. Bloomington: Indiana University Press*. https://doi.org/10.1049/et:20060114.

Jervolino, D. (1990). *The Cogito and Hermeneutics: The Question of the Subject in Ricoeur* (G. Poole, Ed.). London: Kluwer Academic Publishers.

Kaplan, D. M. (2003). *Ricoeur's Critical Theory. SUNY Series in the Philosophy of the Social Sciences*. New York: SUNY Press. https://doi.org/10.1017/CBO9781107415324.004.

Kemp, P. (1996). Ricoeur Between Heidegger and Levinas: Original Affirmation Between Ontological Attestation and Ethical Injunction. In R. Kearney (Ed.), *Paul Ricoeur: The Hermeneutics of Action*. London: SAGE Publications.

Knight, K. (2008). After Tradition?: Heidegger or MacIntyre, Aristotle and Marx. *Analyse & Kritik, 2008*(1), 33–52. Retrieved from http://www.analyse-und-kritik.net/en/2008-1/AK_Knight_2008.pdf.

Lai, W. (1984). Symbolism of Evil in China: The K'ung-chia Myth Analyzed. *History of Religions, 23*(4), 316–343.

Lewin, D. (2012). Ricoeur and the Capability of Modern Technology. In *From Ricoeur to Action*. London: Continuum International Publishing Group.

MacIntyre, A. (2007). *After Virtue: A study in moral theory* (3rd ed.). Notre Dame, Indiana: University of Notre Dame Press. https://doi.org/10.1017/CBO9781107415324.004.

Meretoja, H. (2014). Narrative and Human Existence: Ontology, Epistemology, and Ethics. *New Literary History, 45*(1), 89–109. https://doi.org/10.1353/nlh.2014.0001.

Reijers, W. (2019). Beyond Postphenomenolgy: Ihde's Heidegger and the Problem of Authenticity. *Human Studies, 42*, 601–619.

Ricoeur, P. (1967). *The Symbolism of Evil.* Boston: Beacon Press.

Ricoeur, P. (1971). The Model of the Text: Meaningful Action Considered as a Text. *Hermeneutics and Critical Theory, 38*(1), 316–333.

Ricoeur, P. (1979). The Human Experience of Time and Narrative. *Research in Phenomenology, 9*, 17–34.

Ricoeur, P. (1980). Narrative Time. *Critical Inquiry, 7*(1. On Narrative), 169–190.

Ricoeur, P. (1983a). Action, Story and History: On Re-reading The Human Condition. *Salmagundi, 60*, 60–72.

Ricoeur, P. (1983b). *Time and Narrative – Volume 1* (K. McLaughlin & D. Pellauer, Eds.) (Vol. 91). Chicago: The University of Chicago. https://doi.org/10.2307/1864383.

Ricoeur, P. (1988). *Time and Narrative – Volume 3.* Chicago: The University of Chicago.

Ricoeur, P. (1992). *Oneself as Another* (K. Blamey, Ed.). Chicago: University of Chicago Press.

Ricoeur, P. (2004). *The Rule of Metaphor: The Creation of Meaning in Language.* London: Routledge.

Ricoeur, P. (2014). The Later Wittgenstein and the Later Husserl on Language. *Études Ricoeuriennes / Ricoeur Studies, 5*(1) https://doi.org/10.5195/errs.2014.245.

Romele, A. (2019). *Digital Hermeneutics.* London: Routledge.

Searle, J. R. (2010). *Making the Social World: The Structure of Human Civilization.* Oxford: Oxford University Press. https://doi.org/10.1093/acprof:oso/9780195396171.001.0001.

Shakespeare, W. (1623). *As You Like It* (2005th ed.). San Diego: ICON Group International. Retrieved from http://eprints.kingston.ac.uk/1511/.

Simmel, G. (1900). *The Philosophy of Money* (D. Frisby, Ed.) (3rd ed.). New York: Routledge Classics, 1978.

Van Den Eede, Y. (2010). In Between Us: On the Transparency and Opacity of Technological Mediation. *Foundations of Science, 16*(2–3), 139–159. https://doi.org/10.1007/s10699-010-9190-y.

Verbeek, P.-P. (2005). *What Things Do; Philosophical Reflections on Technology, Agency, and Design.* Pennsylvania: Pennsylvania University Press.

4

A Narrative Theory of Technology

Technical practice, the focal point of technology ethics, has a distinct structure. In using a technology, a practitioner refers through a totality of relevance to that for the sake of which the technical practice exists in social life. To distinguish technical practices that cultivate the virtues from those that do not, we consider those modes of practice in which the practitioner is "awake" and "succeeding". Yet, this raises the question of *genuine* virtuous and responsible conduct: how to distinguish between a criminal hacker who is awake and succeeding in eliciting a zero-day vulnerability to attack a government service, and a responsible software engineer who builds an application that serves the common good? We saw that this question of authenticity leads to the question of temporality, which in turn brings us to narrative as that which endows practices with their temporal significance. Following Ricoeur, we presented a hermeneutic account of technical practice, according to which narrative structures provide for the structural and temporal "wholeness" of a practice: linking basic actions to practices, via life plans, up to the narrative unity of life. This, as Ricoeur explains, provides an answer to the secret of the "nesting of finalities", which we encounter in examining a technical

W. Reijers, M. Coeckelbergh, *Narrative and Technology Ethics*,
https://doi.org/10.1007/978-3-030-60272-7_4

practice, in the movement of ascending complexification and descending specification.

Hence, we have come to understand the crucial role of narrative structures. At the surface level, the impact of narrative and narrative theory on the design and use of technology becomes apparent when we review literature from different disciplines. In the field of computer science, game designers argue that "choices about the design and organisation of game spaces have narratological consequences" (Jenkins 2003, p. 129). Similarly, researchers in artificial intelligence argue that the richness of "narrative intelligence" might solve enigmas resulting from the simplicity of formal understanding of computer intelligence (Mateas and Sengers 1999). The significance of narrativity for social reality becomes clear in discussions about narrative in law and history of technology. It is argued that narrative constitutes an important aspect of legal judgement that goes against purely formalist conceptions of law (Patterson 1990). Moreover, social scientists show that technologies like the Internet are embedded in a "cultural-narrative" (Goodfellow and Hewling 2005) and that historical events like the Fukushima nuclear disaster are understood according to multiple narratives that reveal an interplay of technology, politics and society (Jones et al. 2013). Hence, narrativity appears to be an issue of great interest for studies of technology, the humanities and for the social sciences.

Importantly, narrative structures do not exist in an isolated Cogito, but *externalise* our understanding of technical practices into the world. As Romele argues, Ricoeur's hermeneutics focuses on the way in which our thinking is externalised (Romele 2019, p. 6) through linguistic things: books, reports, movies, and theatre pieces. This is the philosophy of the detour, a philosophy of mediation. Yet, should we assign a privileged role to those "rich" cultural artefacts that most visibly constitute our narrative ecosystem, or does the potential for narrative mediation lie in any artefact that we somehow use of interpret the world? In other words, would narrative mediation be limited to actual texts, or could it be extended to other cultural artefacts like buildings, computers, microwaves, and military drones? Ricoeur (2017) already gave a clue to answering this question, in linking narrativity and architecture. We will follow this path of reasoning to the idea that all technologies, in their multitude of appearance, to a certain extent bear the capacity for narrative mediation. Our claim is therefore that in order to grasp the notion of "technological mediation", insofar as technologies play a constitutive role in our technical practices, we need to investigate technologies' narrative capacity.

In the following sections, we first provide an overview of Ricoeur's narrative theory and its central elements and elaborate on the argument that technologies have narrative capacities. We then construct our account of "narrative technologies", which outlines a theory of technological mediation based on Ricoeur's narrative theory. Finally, we link our narrative account of technological mediation back to the theory of technical practice that we developed in Chap. 3.

Exploring Ricoeur's Narrative Theory

In one of his major works that consists of three volumes, *Time and Narrative*, Ricoeur (1983, 1985, 1988) constructs a comprehensive narrative theory. This theory revolves around a basic model that designates the way in which a narrative, considered according to the paradigm of the text, mediates human reality. This model consists of three conceptual stages that indicate the move from "not having read" to "having read" a narrative. The starting point for the formulation of Ricoeur's theory is the hypothesis that "*time becomes human to the extent that it is articulated through a narrative mode, and narrative attains its full meaning when it becomes a condition of temporal existence*" (Ricoeur 1983, p. 52—emphasis in original). Ricoeur considers human experience, that is irreducibly temporal and social, to be shaped through a *narrative mode.*

The Model of Emplotment

What then, is this "narrative mode"? Ricoeur claims that the *mediation* between time and narrative that is implied in this mode revolves around a process that he designates as *emplotment.* He derives his theory of emplotment from Aristotle's *Poetics* (1902), grounding it in three stages of Aristotle's core concept of *mimesis*, which he characterises as "the active process of imitating or representing something" (Ricoeur 1983, p. 33). Aristotle says about mimesis that "the instinct of imitation is implanted in man from childhood" and that this is an "instinct for 'harmony' and rhythm" (Aristotle 1902, p. 15). Accordingly, the significance of understanding this process of imitation goes beyond the realm of literature and finds its philosophical significance in its purported grounding in human existence in general. Aristotle's definition of mimesis is derived from the overriding principle of *mythos*, which designates the *plot* of a narrative. "The plot is the imitation of the action:—for by the plot I mean here the arrangement of incidents" (Aristotle 1902, p. 25). From this, Ricoeur derives that *emplotment* designates the organisation of events[1] by which people represent action in a plot. Paradigmatic examples are works in the genres of *tragedy* and *comedy* in which characters imitate probable accounts of human actions structured according to a play script or scenario, which is an organisation of events (or "acts").

In a narrative such as a tragedy, the plot *configures* different elements like characters, motivations and events in a meaningful whole. Emplotment, in other words, creates a harmonious concordance out of discordant, heterogeneous elements. Therefore, the movement of a plot accords with a teleological principle, being an "inexorable movement that drives the story toward an anticipated conclusion" (Dowling 2011, p. 6). It makes the reader or spectator say: *this story makes sense*. How then, does emplotment shape the human experience and understanding of temporal existence? Essentially, this amounts to the dialectical movement of *prefigured* time that becomes a *refigured time* through the mediation of *configured time* (Ricoeur 1983, p. 54). To illustrate this: when reading Plato's allegory of the cave our prefigured time involving ideas of human knowledge (before reading) changes by means of interaction with the text in the configured time (during reading) and is subsequently synthesised with our experience of the world in the refigured time (after reading). The reading of Plato's allegory mediates and shapes our experience of the world because it leads us to consider our own experiences as analogous to the shadows cast in the cave in which we are supposed to be imprisoned

[1] Ricoeur explicitly refers to "events" rather than "incidents" because he later in his work concludes that an event as implied in a text might cover a very extensive time-span (Ricoeur 1983, p. 217).

as presented in the allegory. The three phases that constitute the movement of emplotment are defined as follows:

- *Mimesis 1*: the prefigurative phase. This phase consists of the understanding of the world of action: its semantics, its symbolic order and its temporality.
- *Mimesis 2*: the configurative phase. In this phase, characters and events are organised in a meaningful whole, a plot.
- *Mimesis 3*: the refigurative phase of reading. This concludes the narrative circle, of applying narrative to the prefigured world of action.

Mimesis 1 refers to our prior understanding of the human world of action. According to Ricoeur, this prior understanding can be analysed by considering its three basic conditions: human competence to identify action in terms of its *structure*, human competence to identify the *symbolic* mediations of actions, and the human understanding of the *temporal* elements of action. As such, mimesis 1 indicates the initial moment at the beginning of the reading of a text, a moment that is embedded in a social context—in a human "repertoire" from which we engage with new social phenomena. However, mimesis 1 extends beyond the particular text because it concerns the realm of human action in a social setting that is already mediated by narrative. This means that the way we experience our temporal, social existence is in each case shaped by narrative structures (Ricoeur 1983, p. 54). For example, we understand the act of going through the passport control at an airport because we understand the structure of the act (if I'm asked my passport, I hand it over), we understand its symbolic mediations ("EU" desk for EU citizens) and we understand its temporal setting (first I do *x*, then the official does *y*). At the same time, we also understand ourselves, and our life worlds through narratives about our national identities (e.g. "I'm a citizen of the Netherlands), about the economy (e.g. "I lost my job due to the financial crisis"), and about technologies (e.g. "robots are going to render many jobs superfluous"). Thus, whenever we engage with human language, we act from a cultural basis, which means that our understanding is shaped by structures, symbols and temporal elements that are a part of our collective narrative "repertoire" (Ricoeur 1983, p. 64). This view is largely in line with approaches in STS such as Jasanoff's account of socio-technical imaginary, which is conceptualised as a collectively held repertoire of narratives ("visions of the future") embedded in "shared forms of life and social order" (Jasanoff 2015, p. 6). However, it is only the starting point of the model of emplotment, for we are interested foremost in the ways that these "shared forms of life" come to be and are transformed.

Mimesis 2, the phase of narrative configuration, is central in Ricoeur's work and makes explicit how a narrative can configure prefigured time (Ricoeur 1983, p. 64). This phase denotes the process of emplotment: the mediation of prefigured time my means of a *plot*. The plot mediates between individual events and the whole of a story, it brings together heterogeneous factors (characters, goals, interactions) belonging to the realm of action (mimesis$_1$) into a syntagmatic order and it mediates the temporal dimensions of prefigured time. For instance, in the well-known narrative Oedipus Rex several heterogeneous characters (Oedipus' mother, his lover, his father, the king), events (prophecies, trials and murders) and goals (trying to evade the prophecy, aiming to know its truth) are brought together in a syntagmatic[2] narrative whole, which constitutes the story's surprising, but acceptable, resolution (Ricoeur 1983, p. 65). As such, the plot is the organisation of elements of a narrative (characters, events) that makes it possible for someone interacting with a text to follow it to a certain conclusion. Ricoeur elaborates upon the mediation of prefigured time by introducing the idea of two temporal dimensions in the process of emplotment: chronological and non-chronological dimensions of narrative time. The chronological dimension is concerned with an episodic arrangement of events, which characterises the narrative in terms of sequence of events (firstly *this* happened, secondly *that* happened) (Ricoeur 1983, p. 66). The non-chronological dimension of narrative time is concerned with the configuration of events in a meaningful, temporal whole. This implies that the organisation of events is made intelligible, or rather "followable" (referring to the human ability to "follow" a story) in such a way that the order of events leads to a conclusion that characterises the narrative as a whole; that renders its resolution,

[2] "Syntagmatic" refers to the relation between different linguistic elements that occur sequentially in discourse.

according to Ricoeur, acceptable for the reader. It is through the mediation of the non-chronological temporal dimension of a narrative that we can proclaim: *this story makes sense.*

The world of the text and our human world intersect at the moment of refiguration, *mimesis 3* (Ricoeur 1983, p. 71). Refiguration is therefore the third conceptual moment in Ricoeur's model: the moment at which the narrative circle has been closed, or rather completed, and the life-world of the reader is transformed. At this point we should note that Ricoeur acknowledges that the relation between narrative structures and our understanding of the social world is a circular one. That is, the organisation of the narrative structure helps us to understand the social world, but at the same time the understanding of the social world is the basis for any novel narrative structure. To understand this relation we have to deal with a "hermeneutic circle" that consists of the stages of prefigured, configured and refigured time. In opposition to a mere tautology (i.e. "experience *is* narrative structure"), this hermeneutic circle is, in the words of Ricoeur, a "healthy" one, "in which arguments advanced about each side of the problem aid one another" (Ricoeur 1983, p. 76). More specifically, the dimensions of narrative structures that are revealed through textual hermeneutics (e.g. temporal dimensions) are understood in terms of their relation to the world of action, and our engagement with the world of action in turn feeds into the practice of textual hermeneutics. In what follows, we will explore how Ricoeur's model of emplotment can be used to conceptualise the way in which technologies configure human experience and understanding of the social world.

Narrativity and Technological Mediation

How could we employ Ricoeur's narrative theory to understand technological phenomena, related to the practises and outcomes of R&I? To answer this question, we have to explore in what sense technologies can have "narrative" qualities *themselves*, rather than being merely shaped by narratives that are constructed *about* them.

As an initial concern, we face the challenge of reconciling Ricoeur's narrative theory with an understanding of technology. At face value, textual narratives and material technologies seem to be far apart. Consider for instance the apparent gap between literary texts that Ricoeur deals with in the second volume of *Time and Narrative* like Proust's *Remembrance of things past* (Ricoeur 1985) and technologies like cars, computers and cameras. Written stories are made up of words and sentences while cars are made up of materials like plastics and steel. However, we have reasons to question this apparent dichotomy. Ricoeur points at the *generality* of his narrative theory by mentioning a taxonomy of classes of narrative that includes "myths, folklore, fables, novels, epics, tragedies, dramas, films, comic strips, to say nothing of history, painting and conversation" (Ricoeur 1985). Moreover, he claims that the narrative understanding engraved in history also applies to disciplines like "cosmology, geology, and biology" (Ricoeur 1983, p. 135). Thus, his theory of narrative configuration is meant to go beyond the works of literary fiction and historical narrative and to include visual objects like paintings and knowledge in the scientific field of biology. Taking narrative as an ontological concept, as a concept that mediates human experience and understanding in each case, we can see how it does not remain an exclusive concern of the understanding of texts. Arguably, the process of emplotment does not only happen through the interaction with actual written texts, but also through interaction with cultural forms such as theatre plays, movies, paintings, and—as we argue—technologies. Considering the basic notion of emplotment, no absolute separation can be enforced between what one might colloquially want to call "soft" cultural phenomena like texts on the one hand and "hard" technological phenomena on the other hand. Both our understanding of a text and our understanding of a technology are actualised through inter-action. Insofar as interaction is made intelligible through narrative, as we might recall from our discussion of MacIntyre, emplotment governs the meaning of both text-in-reading and technology-in-use. What we additionally argue, alongside Ricoeur, is that narrative understood along the lines of the paradigm of the text can *inform* our understanding of technological mediation, because the process of emplotment is most readily accessible in written texts.

Once we dismiss the strict dichotomy between our understanding of technologies and texts, we open up the possibility of employing the latter to investigate the first. To do so, we need to ask: to what extent can it be said that a technology brings about a "plot" in similar ways as a text does? In other words, the central question for thinking about technologies becomes: can technologies *configure* our narrative understanding in similar ways as written texts do? Kaplan (2006) provides some initial answers to this question. He argues that narrative theory can be used in order to interpret the way in which humans "read" technology (Kaplan 2006, p. 49). Moreover, he points out that there are certain ways in which humans can construct plots to understand technology, for example by telling about the motivations for designing a technology (Kaplan 2002, p. 4). However, taking into considerations the lessons learned from theories of technological mediation, we want to make explicit how humans and technologies co-shape each other by using the concept of narrative technologies. In other words, we argue that humans do not only read technologies, but technologies on the other hand "read" humans, insofar as what is experienced by a user must first be constructed in the technology (Ricoeur 1983, p. 50). Ricoeur defends this argument in a crucial passage:

> "It is the act of reading that accompanies the narrative's configuration and actualises its capacity to be followed. To follow a story is to actualize it by reading it. And if emplotment can be described as an act of judgment and of the productive imagination, it is so insofar as this act is the *joint work of the text and reader*, just as Aristotle said that sensation is the common work of sensing and what is sensed". (Ricoeur 1983, p. 76—emphasis added)

In other words, a reader is only able to imagine and judge the course of events in a narrative insofar as the text simultaneously offers the resources for following and actualising it. Thus, if we take Ricoeur's narrative theory as our guiding light, we need to see using or "reading" a technology as a reciprocal process. Therefore, the term "configuration" as it is used in mimesis 2 seems most suitable here, implying that a configuration of a technology by a human denotes a corresponding configuration of the human by the technology.

To substantiate the move from text to technology (though not substituting the one for the other), let us then briefly reflect upon the characteristics of emplotment that can be found in the design and use of technologies. In order to defend the claim that technologies configure our narrative understanding we need to show that—just as texts—they are involved in the organisation of events. Moreover, we need to defend the claim that the organisation of events includes both chronological and non-chronological dimensions of narrative time. As a starting point, we consider the way technology design is characterised in design literature to ascertain a clearly narrative structure that is used to characterise the functioning of technologies. As an example, we might consider the passage about the "confirmation principle" in the use of dialog boxes in software, as captured by Lidwell et al. (2003, p. 44—emphasis added):

> "Confirmation using a dialog involves establishing a *verbal interaction* with the *person* using the system. It is most commonly *represented* as a *dialog box* on a software display (e.g., "Are you sure you *want* to delete all files?"). In this method, dialog boxes directly ask the *user* if the *action* was *intended* and if they would like to proceed."

In this passage, which serves as a guideline for technology design, heterogeneous factors including characters (people and, more specifically, users), events (action, verbal interaction, representation) and motives (want and intention) are brought together in a coherent whole, in what we might characterise as a plot. This observation is in line with Kaplan's claim that technology design involves narrative. However, we do not only want to show that designers express their ideas in a narrative mode to understand the technologies they create; we want to go further by showing that technologies themselves *configure* a plot. Through reading, a textual narrative actively configures the prefigured understanding that a reader has of her social world (Ricoeur 1983, p. 53). For instance, a reader's understanding of the impact of surveillance technologies might be transformed through a reading of Orwell's *Nineteen Eighty-Four*. Similarly, we maintain that a technology can actively configure a person's narrative understanding of her social world. The interpretation of a technology-in-use entails a coinciding active process of mental "reading" and "writing", understood as two dimensions of the practical activities of writing and reading a text. That is, while we use a technology, we follow its process of emplotment (mimesis 2) and through this our experience and understanding of our social world is transformed (mimesis 3).

In the use of a technology, the technology and its user *together* configure an organisation of events. For instance, we can say that a car, as a technology, configures events such as "starting the engine" and "adjusting the mirrors" in a meaningful whole. Many aspects of this process are related to a prefigured understanding of the world. For instance, before a person has ever driven a car, she will already have an understanding of the way the traffic functions and of the car as a cultural artefact (including understanding of for instance the environmental impact of the use of cars and the impact of driving too fast). Through interaction with the car, however, this prefigured understanding is *configured*; the understanding of both the traffic and the car as an artefact is altered and integrated into a new understanding of the social world.[3] As such, a technology and the technological system in which it is used are actively involved in shaping the way we understand our activities, experiences and relations with other people. The notion that both technology and humans configure an organisation of events is important since it conceptualises, from within a narrative approach, a key insight in contemporary philosophy of technology since Heidegger: technologies are not mere instrumental means, but mediate meaningful human action. Thus, by using Ricoeur's work on narrative we can convincingly support the claim that technologies have the capacity to configure plots, understood as organisations of events.

Narrative Technologies

In this section, we propose a theory of technological mediation that is informed by Ricoeur's narrative theory. While dismissing any strict dichotomy between technical and narrative understanding, our ontological premises are that (1) human experience and understanding are mediated by narrative structures and (2) that these narrative structures are

[3] An existing empirical study about the narratives of older women driving cars clearly shows the dependency of the understanding of driving a car on both the "practical" narratives (narratives about the practice of driving) and "experiential" narratives (narratives arising from the experience of driving) (Siren and Hakamies-Blomqvist 2005.)

generally accessed through involvement with the ready-at-hand, which includes a variety of "things" such as books, movies, houses, computers, and trains. Our theoretical framework offers four central concepts that pertain to technological mediation and are informed by essential distinctions made by Ricoeur in his discussions of textual hermeneutics. To consider technologies in relation to the paradigm of the text, we offer two central concepts that capture distinctions between technologies: *textuality* and *literacy*.[4] "Textuality" pertains to the extent to which technologies approximate the paradigm of the text, whereas "literacy" pertains to the extent to which humans are attuned to involve themselves with the configurations that technologies bring about. To consider the modes of technological configuration that are at stake when interpreting a particular technology, we offer two additional central concepts:[5] *temporality* and *distancing*. "Temporality" pertains to the dimension(s) of time and "distancing" pertains to the abstraction(s) from the world of action that technological configurations bring about.

Textuality, Literacy

Initially, we consider two central concepts that pertain to technologies as particular phenomena that humans encounter in their world—similar to particular instances of narrative such as novels, movies, theatre plays, and so on. Not all technologies actively configure our narrative understanding, even though they are in each case involved in prefigured time. As such, they differ in that they have more or less in common with the paradigm of the text, in terms of their textuality. Furthermore, just as readers have to be literate to a certain extent to read a text, users of technologies have to equally possess a "literacy" to engage in the process of technological emplotment. We use the notion of "literacy" in a broader sense than in its common usage, in which it is characterised as the capacity to read and write in a particular natural language. Here, it is understood (1)

[4] In an earlier paper, "literacy" was instead conceptualised as "interpretative distances" (Reijers and Coeckelbergh 2018).

[5] In an earlier paper, "temporality" was conceptualised as "activity" and "distancing" was conceptualised as "abstraction" (Coeckelbergh and Reijers 2016).

praxis-oriented, extending beyond the mastery of a particular natural language and including the mastery to engage in technological configuration in terms of "coding", "building", "hacking", "designing", "tweeting", and so on, and (2) contextually, meaning that to be literate in for instance using a computer does not only depend on one's own capacities but also on the design of the respective computer, the languages it supports, the user-interface it has, and so on.

First, concerning *textuality*, we need to acknowledge that technologies differ in the extent to which they bring about technological emplotment because some technologies—in particular ICTs—are very similar to the paradigm of the text while others—such as a hammer—are very different from it. The reason for this difference, which will become more explicit in the following sections, can be found in the capacity of these technologies to, through repeated use, transform the experience and understanding of their user's social world. Just as there is a strong difference—though not a discontinuity—between early forms of symbolic communication such as cave paintings and elaborate texts such as Shakespearean plays, and thus a difference concerning the *possibility* of doing textual hermeneutics, there is a difference—though not a discontinuity—between ancient, less textual technologies and modern, more textual technologies. In common words, we could explain this difference by saying: the more a technology-in-use *tells us* something about ourselves and the world we live in, the more it can be regarded as being textual. The notion of textuality assists us in determining the horizon of interpretation, and thereby also postulates a limit for any hermeneutic inquiry of technology. That is, we argue that we can only interpret a technology in use as far as it approaches the paradigm of the text. Consequently, the further a technology is removed from the paradigm of the text, the more we rely on recounted narratives, on testimonies, of its use and significance. This limitation corresponds to the limitation of textual hermeneutics, which as a mode of investigation becomes less feasible the more the phenomenon that is interpreted is removed from the paradigm of the text. The limit that follows from the concepts of textuality is informative insofar as it guides the type of investigation that is appropriate for considering technological mediation: the more a technology is removed from the paradigm of the text, the less we can rely on interpretations of its structure (i.e. design) and the more we rely on recounted accounts of its use.[6]

The extent to which a technology approximates the paradigm of the text is determined by its capacity to actively "emplot" human narrative understanding. Some technologies have little in common with the paradigm of the text and play for the most part a role in our prefigured understanding. For instance, a bridge is predominantly part of a prefigured narrative structure in which events and characters are already configured into a plot: it may be a bridge to transport goods and people across the Rhine River. When a bridge gets built, it plays a role in configuring our narrative understanding (for example by disclosing new areas of a country) but it progressively becomes part of our prefigured time. Such an understanding of technology ties into Heidegger's analysis of the bridge (Heidegger 1977, p. 16): the bridge has become a passive element of human culture in the course of several generations. It configured the narrative understanding of the people who interacted with it once it was built, but in time became an element of their social reality that found *closure*. However, some technologies bring about an *active* process of emplotment of the narrative understanding of their designers and users. Through interaction with these technologies, a process of emplotment is brought about that implies a simultaneous "reading" and "writing" of our narrative understanding. Information and Communication Technologies (ICTs) are exemplary for this type of narrative technologies, being very closely related to the paradigm of the text. This can first of all be derived from their very textual character in a literal sense: many forms of human-computer interaction revolve around mediation by textual information. More importantly, though, ICTs and humans can be said to "co-author" the narratives they engage in. For instance, social media (e.g. Facebook, Twitter) are very text-like technologies, not only with regards to their superficial textual aspects—primarily understood as textual elements of their user-interfaces—, but also, and more importantly, with regards to the configurative capacities of their code-at-work: their capacities to *organise* characters (e.g. avatars) and events (e.g. social media updates) in a

[6] This claim should be understood in relation with an earlier one, which stipulated that we understand action (i.e. technologies in use) through the interpretation of recounted action in narratives.

meaningful whole. This does not refer to the actual reading of the code, like by a software developer, but rather to the narrative structures configured by the code. For instance, interactions with the code of a social media platform can configure a person's narrative structures that form his understanding of practices related to "liking" and "friendship". Thus, instead of looking at the present-at-hand source code in order to investigate the narrative structures it configures, the notion of textuality makes us look at the ready-at-hand plot, the organisation of characters and events, as it is configured by a technology.

Second, with *literacy* we consider the extent to which humans are attuned to engage in a process of technological configuration with particular technologies. As such, we interpret the extent to which a technology and related technological configurations are accessible or inaccessible for different people. To draw an illustrative comparison that assists in understanding this distinction: consider the narrative configuration of George Orwell's *1984* and the commentaries related to this it (e.g. hearsay about the message of the book). A reader of the book might have his ideas transformed concerning the role of technologies in modern society. Someone else, who didn't read the book, might nonetheless comment on it to illustrate something, for instance by referring to a CCTV camera as an "Orwellian technology". In such a situation, the commentator has a prefigured understanding of what an "Orwellian technology" entails but fails to have access to the original process of emplotment (the plot in *1984*). Likewise, there is a difference in understanding of the technological configuration of an ICT between a developer who works with its code (for whom the technological configuration is very accessible), its user who interacts with an application interface and a person who only hears or reads *about* the technology without having used it (for whom the technological configuration is very inaccessible). In contemporary debates about digital technology, this distinction is often characterised in terms of "digital literacy" (Eshet-Alkali 2004), which denotes a difference between the modern "digital natives" who know their way around the technological configurations of digital technologies and the "digital illiterates", to whom the respective technological configurations are inaccessible.

Carr captures this difference by conceptualising two types of narratives as the *practical* "first-order" narrative (narratives configured *by* technologies), and the "second-order" narrative (narratives *about* technologies) that have the same subject but a different aim; namely an aesthetic or cognitive one (Carr 1986, p. 131). This reflects Ricoeur's distinction between "commentary" (which can be a text *about* a narrative) and a narrative configuration itself (Ricoeur 1985, p. 68). Even though both these types of narrative structures have the same form, or in Ricoeur's terms the same *schematism*, and are therefore fundamentally interrelated, we need to distinguish between them because they denote an interpretative distance between humans and technologies that can lead to differences in experiencing and understanding the world. That is, the more accessible a technology is and the more one interacts with it, the more proximate one gets to its first-order narrative configuration. Some technologies, such as hammers or bikes, are accessible to most people, who can be considered "literate" with regards to their capacity to access the first-order technological emplotment of hammers and bikes. Conversely, a layperson who is mildly interested in military drones has much less power in co-authoring their narrative structures and a *different*[7] understanding of them than a core developer of these drones or a military officer operating them. This is not to say that those people interacting with first-order narrative structures *necessarily* have a greater power overall over the technological emplotment of military drones, compared to people who only interact with their second-order narrative structures. For instance, a layperson holding a powerful political position can enforce regulations that strongly influence the prefigured narrative understanding in which the design of military drones takes place. Rather, literacy resonates with what Foucault designates as the authority of expert knowledge (Philipps 1996) as a characteristic of those people interacting with first-order narrative structures of technologies such as military drones. According to this understanding of power, people who are proximate to the first-order narratives are

[7] We explicitly refer to the *difference* in understanding of the narrative structure, rather than for instance to a *lesser* understanding of a layperson. In certain cases, remoteness from technological narrative configuration might actually help inform the human about important hidden aspects of this configuration (think for instance about the position of game addicts vis-à-vis non-addicted non-gamers).

powerful in the sense that they have acquired the capacity to co-author the process of emplotment. A powerful regulator, when being far removed from the first-order narrative, would not have this power to co-author the process of technological emplotment but only to change the prefigured time in which it takes place.

Temporality and Distancing

The following two concepts do *not* pertain to any single technology—considered as a phenomenon comparable to a novel or a theatre play—but to the process of technological configuration. Just as different narrative modes can overlap in a single literary work, such as the modes of historical and fictional narrative, different modes of technological configuration can overlap "in" a singular technology-in-use. The two distinctions we consider are (1) the extent to which a technological configuration brings about a rigid or dynamic *temporality*, and (2) the extent to which technological configuration *distances* itself from the world of action, either engaging with or abstracting from it. Important to note is that the related distinctions between technological configurations are to be considered as capturing differences in degree rather than differences in kind, thereby denouncing for instance the notion that a technology can configure "absolute" abstracting narrative structures.[8] The distinctions function as epistemological "relay stations",[9] constituting the "detour" that Ricoeur envisages throughout his work, with which we can approximate the narrative structures as configured through technical practices. Both distinctions, of temporality and distancing, capture the difference between what Ricoeur conceptualises as the modes of *historical* and *fictional* narrative, which are defined in accordance with the ways they concretise their respective intentionalities (Ricoeur 1988, p. 188). That is, whereas historical narrative aims to achieve a level of strict representation of historical

[8] Ricoeur deals with this "temptation" to consider "absolute" mediations in volume 3 of *Time and Narrative*, in which he attacks Hegel's notion of total mediation between human culture and the individual (Ricoeur 1988, p. 202).

[9] Ricoeur uses the term "relay station" to indicate an epistemological structure which enables one to proceed from one level of a hermeneutic analysis to the other (Ricoeur 1983, p. 182).

events, literary fiction aims at providing both the author of a text as well as its reader with a sense of imaginative freedom, and correspondingly a sense of responsibility (the responsibility of following the plot). We want to show how these features of historical and fictional narrative correspond with the ways in which technological configurations mediate our sense of time and represent reality. Importantly, Ricoeur stresses that the two major narrative modes of history and fiction should not be seen as absolutely separate, but rather as being interwoven (Ricoeur 1988, p. 99). That is, history and fiction borrow from each other's intentionality (the ways in which they are intended to relate to the world) and eventually adhere to the same model of emplotment that makes them intelligible.

First, we consider how technological configuration mediates our understanding of the world by organising the *temporality* of the plot.[10] In line with Ricoeur's notion of narrative configuration, we argue that technological configuration brings about a "third time" (Dowling 2011, p. 76), that is distinct from a purely "cosmic time" (understood as movement of worldly bodies) and a "time of the soul" (understood as a result of the internal flow of consciousness). Dowling illustrates the way in which not only a text, but also a material setting can mediate a human's sense of time. He states anecdotally that "the agonizing wait to get medical attention at the emergency room last week may have taken up three or four hours, but the telling of it"…"may take only ten minutes" (Dowling 2011, p. 47). This observation refers to the notion of a "double temporality" that characterises narrative, of the temporality captured by the narrative itself (e.g. characters in a historical narrative about the First World War that engage in events in the course of four years) and the time of the act of narrating (e.g. finishing the reading of the respective narrative in a single day). That is, there is "a *telos* [end] that carries characters forward in a state of imperfect knowledge about the consequences of their actions, with a narrator who, gazing backward on events from a fixed" … "perspective, has arrived at certain conclusions about their meaning or significance" (Dowling 2011, p. 88—emphasis added). What

[10] In structural terms, we could state that the distinction concerning temporality is concerned with the use of *tenses* (perfect, imperfect, pluperfect, etc.) and the *grammar of time* ("now", "then", "earlier", "today" "twelve o'clock") in narrative discourse.

this teaches us about technological configuration is that the act of *using* a technology (as analogous to the act of reading) does not necessarily coincide with the temporality of the technological configuration that it brings about. For instance, the act of placing a fishing net in a riverbed at dawn might take only a moment, but it involves a narrative structure that stretches throughout the day, including the anticipation of catching some fish before dusk. Similarly, the setting of a calendar entry can be done in a blink of an eye, while it could configure a narrative structure that involves several hours of the day.

How can we comprehend the way in which technological configuration mediates temporality? Ricoeur argues that the organisation of temporality of a narrative depends on two distinct temporal dimensions: a chronological, or episodic, and a non-chronological or configurational dimension (Ricoeur 1983, p. 66). He also refers to these dimensions as the time of the world, public time, and lived time, time of personal experience (Ricoeur 1988, p. 182). The chronological dimension comes about by means of an episodic sequence of events (in common terms: "first this happened, secondly that happened"). This dimension is eventually directed at abolishing the human sense of temporality (Ricoeur 1983, p. 160) by reducing temporal experience to "simple succession". In contrast, the configurational dimension enables a reader to oscillate between the narrative—as a whole—and separate events, to jump between different "times" (e.g. as happens in a flash back), and to create a sense of ending. For technologies that actively configure our social world, this means that they can either enforce a rigid temporal structure on our understanding of the social world or a dynamic one. Consider for example the *car* as a technology. In the practice of driving, some events are determined in a chronological order, like starting the engine *before* driving away. In contrast, adjusting the seat or the mirrors can be done in many different orders; such events are organised according to a non-chronological dimension, involving a sense of personal freedom and responsibility. The non-chronological dimension of configuration also means that a single event only makes sense within the larger whole. For example, adjusting the mirrors refers back to the plot of "driving" as a whole and indeed to the practice-as-narrative: it ties in with the narrative about what good, virtuous driving is, for instance driving includes taking into account and responding to the actions of others.[11] This dimension of narrative time consequently entails a dynamic that closes in on human temporal experience.

We argue that temporality is configured in the process of technological emplotment in two ways: (1) by means of enforcing or relaxing strict successions of events and (2) by means of "connectors", which can be dating mechanisms, version control mechanisms, and tracing mechanisms. The first way is illustrated above by the example of the car and has been applicable to technological configurations throughout human history. The ancient windmill already rigidified temporality in its use by configuring strict successions of events: the brake needed to be removed *before* the wings could be set into motion. However, most of the events configured in the plot of "milling" can be freely organised by the miller, whose use of the mill invokes a similar kind of responsibility as Ricoeur assigns to the reader of a work of fiction: of following the plot. The second way in which temporality is configured relates to what Ricoeur designates as the "connectors" that "make historical time conceivable and manipulable" (Ricoeur 1988, p. 182). Historians use constructions of dating (linking events to dates and clock times), the succession of generations (the replacement of the dead by the living) and traces preserved in documents and archives (marks left by humans and things, indicating a "here" and "now" in the past). For instance, historians refer to traces contained in historical documents such as the Treaty of Versailles, which was signed on the 28th of June 1918, to construct the historical fact of the signing of the Treaty of Versailles. Technologies, and notably modern technologies, configure similar constructions in order to make historical, public time conceivable. Dating happens through the application of dating mechanisms in technological systems, such as the system time in computers. The notion of succession of generations relates to technologies mostly in our prefigured time, to narratives *about* technologies. William Turner majestically illustrated this in his painting "The Fighting 'Téméraire'", in which one of the last battle ships that fought during the Battle of Trafalgar is towed by a paddle-wheel steam tug, symbolising the passing of one

[11] In the next chapter, we will further elaborate on this intersection between a technical practice and narrative.

generation of a technology to another. Yet, in contemporary technologies we see a passing of generations implemented in the form of version control, leading to "old" and "updated" versions of systems. As a figure of speech, this has even been adapted in popular culture, by referring to the predecessors as "1.0" and by referring to the contemporaries as "2.0". Finally, tracing has been implemented in technologies by means of sensing and recording the passing of humans and things, for instance through the sensing and recording of mouse clicks and page views in web browsers.

To illustrate how the two ways of technological configuration of temporality come together, we consider the example of blockchain technologies. A blockchain[12] can be regarded as a public digital ledger (a book of accounts) that contains all the transactions made within its system. "Blocks" are digital, time-stamped records containing the most recent transactions that are cryptographically signed and added to the blockchain in a designated sequence, in a linear, chronological manner (Mcreynolds et al. 2015, p. 3). Whenever a transaction occurs, anywhere in the world, so-called "mining nodes" validate it and add it to the public blockchain, which makes it impossible in principle for the same digital object (which could a monetary transaction, but also a contractual agreement) to be "double spent": to be transacted to different addresses at the same time. This design feature accounts for the so-called "time-stamping problem" by providing "temporal authentication of electronic documents" (Van Rompay et al. 1999, p. 1), and thereby of digital traces. Miners are the agents that collectively control the computational nodes validating transactions within the network. For Bitcoin, the service these miners provide is guaranteed according to a system of incentives, which currently amounts to the miners being rewarded transaction fees and newly created Bitcoins. We can examine the two ways in which temporality is configured in a blockchain technology such as Bitcoin. First, events are organised in a strict succession with limited dynamism. To make a transaction, a user *first* needs to login to her wallet, *secondly* enter a destination address, and so forth. In other words, the process of technological configuration already organises events in the form of a strict succession.

[12] The definition provided only captures the meaning of one particular type of blockchain architecture, namely an open blockchain. Other forms are consortium blockchains and private blockchains.

Second, the Bitcoin network (1) records traces of humans and things that left marks "here" and "now" in the past (for instance: signed transactions), (2) dates these traces by time-stamping them and (3) embeds these operations in a logic of version-control that happens by means of "forks" (updates to the basic code that can lead to the branching of different paths of software development). As such, Bitcoin combines the strict succession of events and the three "connectors" that are also used by historians in historical narrative to make historical, public time conceivable and, as Ricoeur argues, to consequently abolish the human sense of time. It would therefore be no overstatement to characterise blockchain technologies as "history machines".

Second, we propose a distinction between modes of *distancing*, between technological configurations that engage with or abstract from the world of action. Crucial to understanding this distinction is the consideration of the significance of historical narrative on the one hand as "standing for" something that really happened and of fictional narrative on the other hand as instantiating "imaginative variations", concerning events whose actual occurrence is bracketed (Ricoeur 1988, p. 177). We argue that a similar difference in narrative modes (understood as a difference in degree between two extremes) can be applied to the process of technological configuration, meaning that it can (1) configure narrative structures that engage people with the world of action by means of bringing about imaginative variations, or (2) configure narrative structures that are almost strictly representational ("standing for" events that *really* happened) and thereby abstract from the world of action. The first type of technological configuration offers a great sense of freedom (and, correspondingly, of responsibility) by offering the possibility of a multitude of imaginary variations in which the emplotment of characters and events takes place. Conversely, processes of technological configuration that abstract from the world of action fixate our narrative understanding in the form of a public archive containing both *documents* and *traces*[13] (in

[13] Ricoeur argues that it is "the use of documents and archives that makes the trace an actual *operator* of historical time" (Ricoeur 1988, p. 184). In other words, a trace that refers back to something that has been there in the past (such as a fossil, but equally so a validated block on the blockchain that can be "re-traced") combined with the use of documentation or an archive (such as a public ledger that links to digital objects) constitute a sense of historical time.

the form of traceable transactions of digital objects) that stand for, or represent the past "as it really happened". As Ricoeur importantly notes, the term *really* is signified only through the concept "as" ("*as*" it really happened—representing a reality that has itself become inaccessible) (Ricoeur 1988, p. 155). As Ricoeur argues: "this abstraction is a result of *forgetting* the complex interplay of significations that takes place between our expectations directed toward the future and our interpretations oriented toward the past" (Ricoeur 1988, p. 208—emphasis added). This forgetting is the main effect of the abstraction brought about by the "standing for" of a technology's narrative configuration.

We need to make explicit what the above-mentioned process of distancing entails.[14] Note that this second distinction effectively captures the *social* dimension of narrative technologies, for these degrees of abstraction mediate intersubjective (or "being-with-others") relations. Ricoeur argues that the conclusion of a narrative does not need to be predictable but rather acceptable. To defend this claim, he analyses the notion of causal explanation in the paradigm of historical narratives (being narratives about "real" events as opposed to fictional narratives). Although he rejects the idea of historical explanation with recourse to laws, he tries to preserve *causal analysis* and *rational explanation* in history (Ricoeur 1983, p. 128). If we consider explanation of historical facts, Ricoeur argues, we ask for a necessary condition and not for a sufficient, law-like condition. For example, if we ask: "how was it possible that Austria-Hungary declared war on Serbia in 1914?" we might ask for a necessary condition, which was the assassination of Franz Ferdinand. However, the assassination is not a sufficient condition for the declaration of the war. In order to explain historical facts, we need to take into account the teleology that guides the events that make up history, which is grounded in the world of action of individual people. History is concerned with the realm of action, but action placed in a society that has already been configured through narrative activity. For these reasons, we can only explain a historical fact by means of retrodiction (putting past events into a series),

[14] As Ricoeur also suggests, modern technologies render "time" itself abstract: the machines that measure time, such as digital clocks but the blockchain as well, enable an "abstract representation of time" (Ricoeur 1983, p. 63).

not by prediction. This claim is important for our understanding of technological mediation, for it enables us to argue against technological determinism. If technological configurations mediate human narrative time, as we argue, they only provide for necessary reasons and not for sufficient reasons for technologically mediated action. Things could have turned out differently.

On the one hand, technological configurations can engage with the world of action, by organising actual characters and event in a plot. Some types of video games (for instance "open-world role-playing games") offer striking illustrations of this form of configuration, because they allow a player to engage with the world of action of (fictional) characters and events. As such, they configure imaginative variations akin to those configured by works of fiction. On the other hand, Ricoeur explains that narrative structures can be made increasingly abstract by means of construction of so-called quasi-entities that are based on first-order entities, which are actual characters and events (Ricoeur 1983, p. 181). This is done in historical narrative by constructing quasi-entities such as "Germany" and "the Battle of Warsaw", which *stand for* actual characters (e.g. the German minister of foreign affairs) and events (e.g. soldiers performing a charge) and "act" or are "acted in" in ways similar to actual characters and events. For instance, a soldier might be said to participate in "the battle of Warsaw", and "Germany" might be said to have signed the Versailles treaty document. These quasi-entities can be organised in a quasi-plot, which is a plot that is removed from direct interaction of actual characters and events. We argue that similar types of constructions are present in modes of technological configuration. Technologies such as electronic networks and exchanges are quasi-entities that do not directly denote actual people or events. Nonetheless, any attempt aimed at explaining plots that involve these structures will require and activity of referring-back to first order entities: it will require narratives about actual characters that act within a first-order plot. Hence, we argue that abstracting technologies remove themselves from the realm of action by configuring quasi-characters and quasi-events in a quasi-plot (Ricoeur 1983, p. 181).

To illustrate the distinction of distancing in technological configuration, in terms of abstraction from the world of action, we can consider the electronic trading in "derivatives". The construct of a future—a specific type of derivative—was already used in 1730 in Japanese rice markets (Takatsuki 2008) and has evolved into one of the major financial products that are currently traded in the global digital economy (Pryke and Allen 2000). A future is a contract with a price agreement between two parties, based on the buying or selling of an asset at a specific moment in the future. For instance, a rice farmer in Japan might agree with a derivative trader that she is guaranteed to sell his future harvest at a certain price. This allows the trader to bear some of the risk of the rice farmer—the harvest might fail; in which case the income of the farmer is still guaranteed—and at the same time make a profit on it by spreading out her own risk amongst multiple farmers. A derivative is a typical example of a linguistic construct that abstracts from the world of action (the farmer who tries to survive by harvesting from his lands) by constructing quasi-entities (e.g. derivative exchanges, credit risks). These are quasi-entities because similar to quasi-entities in historical narrative they configure our understanding of the world without disclosing their mediation of the world of action. That is, when a local derivative exchange goes down, a referral-back to the world of action needs to be made in order to explain the event (for instance referring to a severe drought that destroyed the harvests of all rice farmers who were securing their livelihood through derivative contracts). As Pryke and Allen argue, in our contemporary world in which derivatives as linguistic contracts have merged with digital technologies, they reflect a cultural shift that is an "outcome of a transformation in our experience of everyday temporal and spatial co-ordinates" (Pryke and Allen 2000, p. 282). Derivatives have become technologically mediated contracts that automatically respond to changes in the quasi-plots they configure (such as price fluctuations, risk indicators), thereby increasingly obscuring the way in which they configure the world of action. However, as the financial crisis in 2008 showed, a referral-back to the world of action, of families losing their income and therefore being unable to pay their mortgage, was necessary to explain how the abstract complexities of derivative trades could have contributed to a global financial catastrophe (Hellwig 2008).

Technical Practice and Technological Emplotment

We now understand how technological configuration mediates practice and thereby how it mediates virtue. Crucial in this understanding will be the notions of *ascending complexification* and *descending specification*, capturing the movement that links practices through technological configuration with more or less distant ideals. Below, we will integrate these concepts into our theory of technological emplotment and show how they relate to its four central concepts of textuality, literacy, temporality and distancing. We will illustrate this by using the example of *driving a car*, incorporating the constitutive rules and ideals comprised by a policy called the "New Driving". This policy was implemented in the Netherlands in 2006 and aimed at mediating the technical practice of driving a car to make it more environmentally friendly (de Goede and Hoedemaeker 2009). The policy was implemented by making students in driving schools aware of the importance of driving sustainably and integrating corresponding practical principles of "the new driving" in their driving lessons. Guiding principles that were adopted in this policy were for instance: (1) switch to a higher gear as soon as possible, and (2) when driving 80 kilometres an hour, drive in the 5th gear. This example is chosen because it is a clear illustration of the twofold movement that links basic actions to more or less distant ideals.

In conceptualising a particular practice, *textuality* shows us *where to look*. That is, when considering technical objects that are less textual, such as a bridge, we need to focus on the narratives recounted *about* a particular object to understand how it mediates our practices (mimesis 1). For instance, one could recount how a particular bridge has been crucial for starting a transportation company that delivers goods coming from one city to another city. Such a narrative tells us more about how the bridge mediates particular practices than about the design of the bridge as such. A car is different, in that its design to a much greater extent co-authors the narrative structures that mediate the practice of driving (mimesis 2). In our example, the gearbox in a car enables certain basic actions that can be linked in an action-chain, for instance shifting back gears when closing in on a junction with red traffic lights. These actions are understood according to certain constitutive rules that are considered to be standards of excellence, for instance the rule: switching down gears and pressing the brake in front of a traffic light counts as a valid response in traffic. The practice of driving a car links these basic actions and action-chains with more or less distant ideals, such as that of sustainability. Or another example: driving a car can be linked with the life plan of being a taxi driver, in which case the technical practice of driving is nested with regard to the narrative of a person's professional life.

Literacy, then, shows us *at whom to look*, and to whom a particular process of technological configuration is accessible. The technical practice of driving is remarkable in this regard because it is widely shared across people from different walks of life. As such, it is one of the few technical practices that is relatively difficult to master but that is nonetheless mastered by many people in society with varying capacities and dispositions. When investigating the technical configuration of driving a car, we therefore do not have to limit our discussions to the authority of expert knowledge. Instead, we can consider that the notion of *driving virtuously* is accessible to a vast share of the population. The nesting of finalities involved in the technical practice of driving is accessible to all those who are able to drive a car. In other words, almost everyone can become a virtuous driver, whereas for instance only very few people can aspire to be virtuous nuclear scientists. This also means that the notion of driving responsibly is one that applies to each driver on the road. Practically speaking, literacy is an important aspect to consider in technology ethics when discussing which stakeholders are *relevant* to be included in the process of practising ethics in research and innovation contexts.

Earlier in this chapter, we already invoked the technical practice of driving a car when considering the *temporality* involved in the process of technological configuration. We argued that the design of the car configures both the chronological and non-chronological dimensions of time in the practice of driving, the latter of which provides the driver with a sense of freedom and responsibility to follow corresponding plots. Concerning "shifting gears", as being nested with regards to the practice of driving, we can see how the manual gear system provides more freedom and responsibility when contrasted with automatic gear systems. The driver is free to determine the order of basic actions and can make mistakes when attempting to apply the technique, but also—and more importantly—with regards to "being awake and succeeding" while being on the road. For instance, shifting gears becomes an integral aspect of responding to a sudden event in a traffic situation, such as the appearance of a speeding car at the rear, which inclines a driver who is awake and succeeding to keep driving in lower gears for a while to catch up speed quickly enough. This is accomplished in the practising of driving according to the movement of *ascending complexification*, because it links basic actions of switching gears with the ideal of being a virtuous driver, who is resolved to act in a definite moment. In other words, the a-chronological dimension of time in the technological configuration of a car with a manual gearbox allows for the action of shifting gears to be linked with the notion of being a virtuous driver.

A similar interpretation follows from the consideration of *distancing* in the technological configuration of a car, whereas the plot remains proximate to the world of action: one engages in driving in close proximity to other people driving their cars and other vehicles in the environing traffic. Self-driving cars would constitute a paradigm shift in this regard, because the models on which their responses would depend abstract from the world of action by linking generalised entities (quasi-characters) with generalised patterns of movement (quasi-events). In the case of a car with a manual gearbox, though, we argue that because the technological configuration is proximate to the world of action, it can bring about imaginative variations. As such, a car provides a "personal" understanding and experience of the practice of driving: because the technology configures basic actions such as shifting gears proximate to the world of action, it allows for a driver to link them with ideals such as those captured by a "driving style". One can link one's basic actions to being a "sportive driver", but also a "road devil". Hence, this type of technological configuration that approximates the world of action both enables virtue and vice.

The policy of "New Driving", which promotes environmentally friendly driving, discloses the opposite movement, one of *descending specification*, for it links remote ideals of environmentally friendly life with action-chains and basic actions. That is, these remote ideals translate into policy that governs the technical practice of driving, which accordingly translates into changes in the constitutive rules that mediate basic actions. Again, technological configuration is important to consider, for it allows for this particular movement of descending specification to unfold. Concerning the *temporality* of technological configuration, the sense of freedom and responsibility brought about by the a-chronological dimension of time allows drivers to both be aware of the environmental impact of their technical practice and inclined to do something about it. In other words, it links the in-order-to of shifting gears with the for-the-sake-of-which of living sustainably in a political community. A similar interpretation can be made with regards to *distancing*: the close proximity to the world of action allows drivers to evaluate their driving style and the driving styles of others according to the basic actions that link to the ideals of a sustainable life.

Conclusion

In this chapter, we used Ricoeur's narrative theory to construct a theory of technological mediation. We established four hermeneutical concepts that make explicit how technological emplotment mediates our social world (see Table 4.1 below). First, we considered *textuality*, which denotes a distinction between technologies that are passive elements of prefigured time and technologies that actively configure characters and events in a plot. Second, we considered *literacy*, which denotes a distinction between technologies that are accessible, meaning that many people have access to their first-order emplotment, and technologies that are inaccessible. Third, we considered *temporality*, which denotes a distinction between technological configurations of dynamic, lived time, and of rigid, public time. Fourth, we considered *distancing*, which denotes a distinction between technological configurations that engage

Table 4.1 An overview of (1) the four central concepts of our theory of narrative technologies, (2) the distinctions between technologies and modes of technological configuration denoted by these concepts and (3) particular technologies that illustrate extremes captured by the distinctions, which are also used in the text

Concept	Distinction	Illustration
Textuality	Between technologies that actively emplot narrative understanding and that become and remain passive.	Social media actively emplots narrative understanding, while a bridge becomes a passive element of prefigured narrative understanding.
Literacy	Between technologies that are accessible and that are inaccessible.	A hammer is accessible, while a military drone is inaccessible.
Temporality	Between technological configurations that bring about a rigid, public sense of time, and a dynamic sense of time.	A blockchain technology configures the rigid (public) dimension of time, while a windmill configures the dynamic dimension of time.
Distancing	Between technological configurations that engage the world of action, and abstract from it.	A video game engages the world of action, while a derivative trading system abstracts from it.

with and abstract from the world of action. Before we proceed to the next chapter, it is important to stress that no direct *normative* judgement can be derived from the hermeneutic distinctions we established. Each type of technology and technological configuration has to be understood within the frame of a certain technical practice. For instance, for certain monetary practices, it is beneficial to have processes of technological configuration that abstract from the world of action, while these same processes might be detrimental to practises in care-settings. In the next chapter, therefore, we will embed the theory of narrative technologies in a theory of technical practice that will allow us to draw a connection with the virtue ethics of technology approach as established in the previous chapter.

However, even though we have acquired a much clearer understanding of the structure of technical practices and the role of technologies in mediating this structure, we have not yet solved our main dilemma. That is, we have not yet come to the point of distinguishing between a technical practice that *genuinely* cultivates the virtues and one that does not. For this, we will have to investigate the link between narrative technology and ethics. We will have to show how the narrative structures involved in technological emplotment give rise to the relevant type of *ethical* human activity, namely technical practice that cultivates the virtues. The difference between narrative and ethics lies in the problematic of making normative distinctions, which entails that the narrative mode can go to extremes where sameness and selfhood can be negated. Narrative structures can make us doubt that there is any ground to the question "who am I?" and consequently any ground to people's moral evaluations and estimations. Ricoeur shows that concrete practice intervenes at this point and binds narrative to our everyday existence, stating that the pragmatic answer to the doubt engendered by narrative is "here is where I stand!" And by means of this practical testimony, praxis leads us back towards the ethical aim.

With this difficulty in mind, the next chapter will work out the tripartite structure of technical practice, narrative, and ethics. It will do so by first connecting our endeavour with the tradition of virtue ethics, and especially with the recent attempt by Shannon Vallor to establish virtue ethics in the field of technology ethics. It then will turn back to Ricoeur, and notably to his work on his "Little Ethics", which will establish the *ethical aim* as an answer to the main dilemma of technical practice.

References

Aristotle. (1902). *Poetics* (S. H. Butcher, Ed.). (The Poetic). London: Macmillan and Co.

Carr, D. (1986). Narrative and the Real World. *History and Theory, 25*(2), 117–131. https://doi.org/10.2307/369249.

Coeckelbergh, M., & Reijers, W. (2016). Narrative Technologies: A Philosophical Investigation of the Narrative Capacities of Technologies. *Human Studies, 39*(3), 325–346. https://doi.org/10.1007/s10746-016-9383-7.

de Goede, M., & Hoedemaeker, M. (2009). Daadwerkelijk rijgedrag met Het Nieuwe Rijden (No. TNO-DV 2009 C284).

Dowling, W. C. (2011). *Ricoeur on Time and Narrative: An Introduction to Temps et récit*. Notre Dame, IN: University of Notre Dame Press.

Eshet-Alkali, Y. (2004). Digital Literacy: A Conceptual Framework for Survival Skills in the Digital Era. *Educational Multimedia and Hypermedia, 13*(1), 93–106.

Goodfellow, R., & Hewling, A. (2005). Re-conceptualising Culture in Virtual Learning Environments: From an "essentialist" to a "negotiated" Perspective. *E-Learning and Digital Media, 2*(4), 355–367. https://doi.org/10.2304/elea.2005.2.4.355.

Heidegger, M. (1977). *The Question Concerning Technology*. New York: Garland Publishing.

Hellwig, M. (2008). Systemic Risk in the Financial Sector: An Analysis of the Subprime-Mortgage Financial Crisis. *MPI Collective Goods Preprint, 2008*(43), 77. https://doi.org/10.1007/s10645-009-9110-0.

Jasanoff, S. (2015). Future Imperfect: Science, Technology and the Imaginations of Modernity. In S. Jasanoff & S. Kim (Eds.), *Sociotechnical Imaginaries and the Fabrication of Power* (Vol. 1, pp. 1–49) https://doi.org/10.1017/CBO9781107415324.004.

Jenkins, H. (2003). Game Design as Narrative Architecture. *Response, 44*(3), 1–15.

Jones, C. F., Loh, S.-L., & Sato, K. (2013). Narrating Fukushima: Scales of a Nuclear Meltdown. *East Asian Science, Technology and Society, 7*(4), 601–623. https://doi.org/10.1215/18752160-2392860.

Kaplan, D. M. (2002). *The Story of Technology*. Retrieved from http://www.pages.drexel.edu/~pa34/TheStory of Technology.pdf.

Kaplan, D. M. (2006). Paul Ricoeur and the Philosophy of Technology. *Journal of French and Francophone Philosophy, 16*(1/2), 42–56.

Lidwell, W., Holden, K., & Butler, J. (2003). *Universal Principles of Design*. Gloucester: Rockport Publishers. https://doi.org/10.1007/s11423-007-9036-7.

Mateas, M., & Sengers, P. (1999). Narrative Intelligence. In *Proceedings AAAI Fall Symposium on Narrative Intelligence* (pp. 1–10) https://doi.org/10.1016/S0890-4065(99)80003-6.

Mcreynolds, E., Lerner, A., Scott, W., Roesner, F., & Kohno, T. (2015). Cryptographic Currencies from a Tech-Policy Perspective: Policy Issues and Technical Directions. In *In Financial Cryptography and Data Security: FC 2015 International Workshops, BITCOIN, WAHC, and Wearable. Vol. 8976*. San Juan: Springer.

Patterson, D. M. (1990). Law's pragmatism: Law as practice & narrative. *Virginia Law Review, 76*(5), 937–996.

Philipps, L. (1996). Discursive Deficits: A Feminist Perspective on the Power of Technical Knowledge in Fiscal Law and Policy. *Canadian Journal of Law & Society, 11*, 141–176. https://doi.org/10.1017/CBO9780511609800.010.

Pryke, M., & Allen, J. (2000). Monetized Time-Space: Derivatives – Money's "new imaginary"? *Economy and Society, 29*(2), 264–284. https://doi.org/10.1080/030851400360497.

Reijers, W., & Coeckelbergh, M. (2018). The Blockchain as a Narrative Technology: Investigating the Social Ontology and Normative Configurations of Cryptocurrencies. *Philosophy & Technology, 31*(1), 103–130. https://doi.org/10.1007/s13347-016-0239-x.

Ricoeur, P. (1983). *Time and Narrative – Volume 1* (K. McLaughlin & D. Pellauer, Eds.). Chicago: The University of Chicago. https://doi.org/10.2307/1864383.

Ricoeur, P. (1985). *Time and Narrative – Volume 2* (K. McLaughlin & D. Pellauer, Eds.). Chicago: The University of Chicago.

Ricoeur, P. (1988). *Time and Narrative – Volume 3*. Chicago: The University of Chicago.

Ricoeur, P. (1992). *Oneself as Another* (K. Blamey, Ed.). Chicago: University of Chicago Press.

Ricoeur, P. (2017). Architecture et narrativité. *Études Ricoeuriennes / Ricoeur Studies, 7*(2), 20–30. https://doi.org/10.5195/errs.2016.377.

Romele, A. (2019). *Digital Hermeneutics. Digital Hermeneutics*. London: Routledge. https://doi.org/10.4324/9780429331893.

Siren, A., & Hakamies-Blomqvist, L. (2005). Sense and Sensibility: A Narrative Study of Older Women's Car Driving. *Transportation Research Part F: Traffic Psychology and Behaviour, 8*(3), 213–228. https://doi.org/10.1016/j.trf.2005.04.008.

Takatsuki, Y. (2008). The Formation of an Efficient Market in Tokugawa Japan. *ISS Discussion Paper Series, F-143*.

Van Rompay, B., Preneel, B., & Vandewalle, J. (1999). The Digital Timestamping Problem. In A. Barb, E. C. van der Meulen, & P. Vanroose (Eds.), *Proceedings of the 20th Symposium on Information Theory in the Benelux* (pp. 71–78). Enschede: Werkgemeenschap voor Informatie- en Communicatietheorie.

5

Narrative Ethics of Technical Practice

A major difficulty in a virtue ethics of technology is to find a ground for distinguishing between genuinely "good" technical practices that cultivate the virtues and technical practices that do not. In the previous chapters, we have found certain clues as to what this distinction might entail. First, good technical practices are ones that are engaged in seriously, whereas other technical practices—which are not necessarily "bad", because, in fact, they are often needed too for activities of training—allow for practitioners to err and to be "sleepy". As such, an apprentice roboticist is expected to tinker with parts and circuits, sometimes making robots that do not work or fully go "out of hand", whereas a nuclear scientist working on a new power reactor is expected to act seriously and succeed in each case, not to produce any risk for nuclear disaster. What we can observe from this, is that with the possibility of virtue comes the possibility of vice, and the need for responsible conduct. Second, we found out that this distinction can be expressed according to a structure of practice, one that involves the relations of "in-order-to" and "for-the-sake-of-which". That is, serious technical practices involve being awake and succeeding in linking the in-order-to of the practice (e.g., programming the robot in order to make it navigate the room) to a

W. Reijers, M. Coeckelbergh, *Narrative and Technology Ethics*,
https://doi.org/10.1007/978-3-030-60272-7_5

for-the-sake-of-which of this Dasein in the political community (e.g., for the sake of keeping the house clean). Third, this led us to the problem of the nesting of finalities, which needs to be addressed to distinguish between those serious technical practices that are genuinely good and those that might actually be vicious (think of a "good" assassin).

In a way, the outlining of a theory of technical practice that is linked to a narrative theory of technical mediation has been a lengthy response to the problem of the nesting of finalities. We came to understand that technologies mediate narrative structures that in turn mediate technical practices through the movement of ascending complexification and descending specification. As such, technologies mediate the way basic actions are linked practices and to life plans, which in turn are linked to the narrative unity of life. This narrative unity, however, is not absolutely whole, but always incomplete; yet moving towards completion. As such, it constitutes an unstable mixture of fabulation and actual experience. A hermeneutic circle persists between the narrative structures that mediate our understanding and actual experience from which new narrative structures arise. The instability that a narrative account of technical practice leaves us with, has to be stabilised to gain a better understanding of those technical practices that genuinely cultivate the virtues, because they link to what Ricoeur designates as the *ethical aim*; a concept that links the description, via narrative mediation, to *prescription* of technical practices. Hence, the hermeneutic circle is revisited at the ethical plain.

It is at this point that we need to turn from narrative to ethics. We will not straight away initiate some new ethical theory from scratch, but start from recent work on the relation between technical practice and virtue. In the following sections, we first provide an overview of virtue ethics in general, which builds on the work we already did on MacIntyre in Chap. 2, but then moves to a more specific discussion of recent work on virtue ethics of technology (Vallor). We then formulate a critique of this work, appraising the way in which it has introduced the concern of virtue in technology ethics but problematising its lack of engagement with the phenomenon of technological mediation. This, in turn, leads us to consider Ricoeur's notion of the ethical aim in his 'little ethics', which we reinterpret in light of the theory of technical practice.

Virtue Ethics and Technology

In this section, we briefly explore virtue ethics in general and how it has recently been introduced to the field of technology ethics. Virtue ethics builds on the idea that human agents possess certain consistent dispositions or states of character, called the virtues, which inclines them to engage in right action in a particular context. As an approach in ethics, it does not primarily focus on the evaluation of particular actions, but on evaluating the agent's character. The seminal paper that has been responsible for the revival of Aristotelian virtue ethics in the (English speaking) Western world is G.E.M. Anscombe's *Modern Moral Philosophy*, in which the author states that when we have the outlines of a philosophy of psychology, in which notions such as "action, intention, wanting and pleasure" can be accounted for, we might *eventually* start advancing the concept of "virtue" as the basis for a form of ethics (Anscombe 1958, p. 12). Even though Anscombe did not herself present a theory of virtue ethics and was reluctant to adopt the Aristotelian system due to its perceived lack of clarity (notably regarding concepts such as eudaimonia or human flourishing), a novel approach in moral philosophy emerged that started to develop the Aristotelian notion of virtue, with Philippa Foot (1978), John McDowell (1979), Rosalind Hursthouse (1999), and Alasdair MacIntyre (2007) as its main proponents. What these scholars have in common is not only their focus on virtue, but—in line with Anscombe—also a strong disagreement with the established Kantian and utilitarian traditions in moral philosophy. Yet, Martha Nussbaum (1999) questions whether virtue ethics genuinely offers a *distinct*, or incommensurable approach that can be contrasted with Kantian ethics and utilitarianism, because thinkers in these traditions also assign importance to the notion of virtue in their theories. We tentatively agree with her assessment, but also argue that virtue ethics at least presents a *refocusing* or correction of modern moral philosophy by a return to the notion of virtue as developed in ancient Greek ethical theory. Accordingly, a notable difference between virtue ethics and other approaches in moral philosophy is the *emphasis* on the agent as virtuous or vicious rather than on the rightness or wrongness of an action.

In the literature on virtue ethics, the following central questions are posited: (1) *what* is virtue; (2) *which* (cardinal or principal) virtues can we distinguish; and (3) *how* are virtues "cultivated"? A human being has certain dispositions that we commonly say belong to her "character". Some of these can be regarded as "general beneficial characteristics" (Foot 1978, p. 2), which can be conceptualised as "virtues". It is the possessing of virtues, or rather of a certain kind of "virtuous being" that determines the goodness of a moral agent. The virtues themselves are socially cultivated, rather than naturally acquired, but do depend on a teleological notion of the good life for human beings (the human *ergon*, or "function") (NE, 1.7, 1098a15–20, trans. Irwin). This teleological notion is captured by the concept of *eudaimonia*, which cannot be easily translated into the more familiar concept of happiness (which might be understood as a pleasurable state), but rather designates the end of a human life as a whole that needs no qualification. While other ends in a human life, such as health, can have a qualification (e.g., we are healthy to live in accord with eudaimonia), eudaimonia does not have any further qualification (e.g., it does not make sense to state that we live in accord with eudaimonia to be healthy). From an Aristotelian point of view, it seems to be the ultimate end of human life. Accordingly, a virtue needs to be a disposition that is acted upon in accordance with eudaimonia and that *endures* rather than one that might be deployed in a single action or that may lay dormant in an individual. An agent can therefore be assessed according to an evaluation of enduring dispositions to act in certain way in particular circumstances and not according to an evaluation of a single action (e.g., one apparently kind act does not make its agent kind). Different from Kantian ethics and utilitarianism, which argue from the idea of ethical principles towards an idea of the good life, virtue ethics holds that acting on the virtues in line with eudaimonia yield rules of action (Hursthouse 1999, p. 39). For instance, virtue ethics maintains that by engaging in just practices we become just people and therefore acquire a notion of rules that just practices adhere to (e.g., a sense of procedural justice).

Based on a certain conception of what virtue is, distinct versions of virtue ethics offer different heuristics or lists of virtues. MacIntyre discusses types of these heuristics, such as the lists of Christian virtues, Homeric virtues and Benjamin Franklin's utilitarian virtues (MacIntyre

2007, p. 185). A heuristic in a certain version of virtue ethics stands for a distinct, systematic theory of virtue (MacIntyre 2007, p. 183). Aristotle articulated the most influential list of virtues in Western philosophy. He distinguished between virtues of character and virtues of thought, which relate to one-another. Virtues of character belong to the part of the soul that is not rational itself but nonetheless can "share in reason" (NE, 1.13, 1102a14–18, trans. Irwin). In other words, these are virtues that allow a person to have certain emotional dispositions or rather *feelings* such as fear, confidence, pleasure and pain, in the right way that is mediated by the right reasons. A way to understand such a virtue is by considering the mean within two types of emotional extremes: of excess and of deficiency. For instance, an excess of fear leads to cowardry while a deficiency of fear leads to recklessness. The virtue of courage represents the mean disposition between these extremes and is mediated by rational deliberation—notably by the intellectual virtue of *phronesis* or prudence. Amongst the virtues of character, we can count courage, temperance, generosity, magnanimity, mildness, truthfulness, wit, friendship and justice. Virtues of thought are dispositions to act that are fully rational (NE, 1.13, 1103a1–5), belonging to the human intellect. In other words, these virtues do not mediate our feelings, but aspects of reason. The virtues of thought are craft knowledge, scientific knowledge, prudence, wisdom and understanding. Prudence, or *phronesis*, which is a disposition to grasp the truth concerned with (a particular) action about what is good or bad for the human function, takes a central position because it connects all the virtues by linking practice with reflection. As such, prudence is a required virtue for all the other virtues to be cultivated and acted upon.

Additionally, theories in virtue ethics make explicit *how* virtues are cultivated, or how a human being becomes virtuous. MacIntyre is perhaps the scholar who has constructed the most widely adopted theoretical basis for answering this question. In Chap. 2, we already came across his theory of practice; here we will reiterate this theory in light of its relation to virtue. MacIntyre takes the position, one that we will adopt hereafter, that the "how" question is the one that unifies different, seemingly incommensurable theories of virtue (MacIntyre 2007, p. 186). Virtue, for MacIntyre, is "an acquired human quality the possession and exercise of which tends to enable us to achieve those goods which are internal to practices and the lack of which effectively prevents us from achieving any such goods" (MacIntyre 2007, p. 191). He argues that to understand the logical development of the concept of virtue, we need to pay attention to the different *stages* in which virtues are cultivated. First, there is the stage of *practice*, in which a human being needs to engage in order to cultivate her virtuous being. MacIntyre explains that a practice should be understood as a "socially established human activity through which goods internal to that activity are realised" (MacIntyre 2007, p. 187). Goods internal to a practice are those related to the desire or wish to excel or improve in the respective practice. An example would be practicing a musical instrument in virtue of improving the mastery of playing it. As such, internal goods are opposed to external goods, which are reasons for engaging in an activity that are external to it. An example would be working to acquire money, because a monetary reward is external to the actual activity of work. The notion of practice used by MacIntyre relates to one of the three senses of *praxis* that Irwin discusses in his authoritative translation of Aristotle's *Nicomachean Ethics* (Aristotle 1999, p. 315). First, praxis can be used broadly for all voluntary, intentional actions. This type of praxis includes those actions performed to harness the necessities in life (e.g., nutrients). In its second, narrower sense, praxis can be rational action based on a decision. This relates to the realm of crafts, or *technê*. Finally, praxis can have a third, most narrow meaning of rational action, which is an end in its own right. It is such instances of praxis, designated as "practices" by MacIntyre, that enable an agent to cultivate her or his virtues. Before, in Chap. 2, we discussed how MacIntyre's theory of practice in a way similar to Ricoeur's includes notions of narrative (through which practices are made intelligible), and moral tradition, as an extended narrative argument that grounds moral communities.

So far, we have gained three things from the discussion. First, a virtue can be characterised as a disposition to act in accordance with a sense of completion of a human being, called eudaimonia. Second, even though virtue as such might be one, in normative discourse a heuristic of virtue is employed that distinguishes between virtues of character and of thought. Third, virtues are cultivated through practices, and by extension through technical practices. Yet, we have not yet encountered *technology* in an explicit way. The following section turns to technology and explores the most influential attempt yet to introduce the notion of virtue in technology ethics.

Technology and the Virtues

Shannon Vallor, in *Technology and the Virtues* (2016), offers a first comprehensive ethical theory of technology that adopts the virtue ethics approach. We will examine this approach, to ascertain to what extent it might offer us a conception of the "ethical aim" that would stabilise the notion of a virtuous technical practice. Of the earlier mentioned scholars (Anscombe, Foot, McDowell, Hursthouse and MacIntyre) in virtue ethics, Vallor's framework seems to be most closely related to that of MacIntyre, since she argues that the main aim of her book is to offer "an ethical *strategy* for cultivating the type of moral character that can aid us in coping, and even flourishing" (Vallor 2016, p. 10—original emphasis) under the challenging conditions of emerging technologies. MacIntyre's framework is also the only one next to Aristotelian, Buddhist and Confucian frameworks that she explicitly acknowledges and adopts. To arrive at this point, the leading question of Vallor's work is to figure out "what we will *do* with these technologies once we have them, and what they will do with *us*" (Vallor 2016, p. 5—orginal emphasis).

In constructing her theory of virtue ethics of technology, Vallor argues that she contributes to the existing virtue ethics tradition in three significant ways. First, she strengthens the criticism raised against utilitarian and Kantian ethics by conceptualising "acute technosocial opacity" (Vallor 2016, p. 6) that renders us unable to make decisions based on universal rules or foreseen consequences of technologically mediated actions. Second, she constructs her framework by drawing from Aristotelian, Confucian and Buddhist ethics, by which she manages to propose a virtue ethics of technology that can potentially be embraced by different cultures—perhaps even globally—and is therefore less affected by the presumed Western bias that a purely Aristotelian ethics. Because of the global character of emerging technologies, most notably global information and communication technologies, this is an important aspect of

Vallor's theory and further developing it into a global virtue ethics of technology would be quite an achievement. Third, she offers a detailed account of the practice of cultivation of virtue and a list of "technomoral"[1] virtues, both of which are responsive to the challenges posed by contemporary emerging technologies.

Vallor presents seven "core elements" or perhaps rather *conditions* of the practices that cultivate the so-called "technomoral" virtues.[2] A practice would need to fulfil each condition in order for it to cultivate the virtues. Together, these conditions could provide us with something like an ethical aim. First, there are four conditions that enable practical wisdom or prudence. Practices should be engaged with habitually, meaning that one repeats them, guided by moral exemplars and eventually gets accustomed to them in a positive sense (Vallor 2016, p. 74). In line with MacIntyre's account, practitioners should also understand practices in the context of their relations with other members of the moral community, making them responsive to the particular aspects of these relations (Vallor 2016, p. 83). The agent while engaged in practice should additionally reflect on her weaknesses and faults, both pertaining to the generality of the human being as well as to her particular character (Vallor 2016, p. 90). Then, in resonance with Aristotle, agents should position their practices within a moral will to attain moral goodness (or rather, to keep striving for eudaimonia) (Vallor 2016, p. 97).

Second, having enabled a state of practical wisdom, practitioners also need to get to a state of completion (telos), which requires three additional conditions. Practices require in particular a habit of "moral attention," meaning that the agent is attentive to problematic moral situations both in a cognitive sense (knowing why a situation is morally

[1] We will not adopt Vallor's notion of "technomoral virtues" but refer to "virtues" instead, since we argue that virtues are in each case cultivated by practices that are mediated by technologies. This renders the adjective "*techno*moral" superfluous. Additionally, the emphasis on the term "moral" introduces the problematic of this term's connotation with "obligation" and "cultural norms", which disregards the distinction between ethics and morality, as discussed for instance by Ricoeur (1992, p. 170).

[2] Our presentation of the seven core elements of the practices for the cultivation of technomoral virtue is a very brief summary of the lengthy discussion presented by Vallor. Therefore, some important aspects of each of these core elements might be missing in our account. However, the general picture arising from this discussion should at least provide for a reliable overview of the aspects of each core element that are crucial for our later analysis.

problematic) and through engaging in an appropriate emotional response (Vallor 2016, p. 104). Moreover, an agent has to be able to choose well amongst a range of alternative courses of action given a particular situation (Vallor 2016, p. 109). Finally, practices require an appropriate extension of moral concern: meaning that virtuous practices should be directed at "the right beings, at the right time, to the right degree, and in the right manner" (Vallor 2016, p. 117).

Having dealt with the *how* of the cultivation of virtue, Vallor proceeds with presenting a heuristic of virtues that she argues answers the *what* of "technomoral" virtue. Rather than delving into the ontology of virtue, she answers a specific type of "what" question, namely "what, or which virtues can we distinguish?" In answering this question, Vallor draws from the three virtue ethics traditions she includes in her approach, Aristotelian, Confucian and Buddhist, and synthesises them within the context of a reflection on the impacts of emerging technologies. The following list of virtues results from her deliberation (Curzer 2017; Vallor 2016). We present briefly each virtue's name, and explication in relation to technology ethics.

1. *Honesty*; related to need to address the changing nature of interpersonal communication through digital media.
2. *Self-control*; related to the need to deal with the increasing range of temptations offered through digital media.
3. *Humility*; related to the need to balance our tendency to under- or over-estimate the benefits and harms of emerging technologies.
4. *Justice*; related to the need to address the increasing unjust distribution of resources and power through technological communication channels.
5. *Courage*; related to the need to overcome the challenges humans face of existential proportions (e.g., climate change) that correspond with technological progress.
6. *Empathy*; related to the need to deal with the influence of digital culture on our interpersonal relations (e.g., violent discussions on social media).
7. *Care*; related to the need to properly balance impacts of technologies on human intimacy (e.g., through the introduction of care robots).

8. *Civility*; related to need to respond to individualisation that comes along with emerging technologies and the neglect of public life.
9. *Flexibility*; related to the interdependence between moral cultures caused by technologies.
10. *Perspective*; related to the need for a better understanding of the impacts of emerging technologies.
11. *Magnanimity*; related to the need for leadership in dealing with the impacts of technologies.
12. *Wisdom*; related to the need to be knowledgeable of the technological world in dealing with its impacts.

Vallor shows how these 12 virtues are particularly relevant in the twenty-first century, in a context of increasingly ubiquitous impacts of technologies. However, as Curzer indicates, she less convincingly defends these virtues as somehow typically "technomoral", or uniquely applicable to our current technological condition (Curzer 2017). Even though these virtues are *relevant*, they might not be *typical* for our current age. Moreover, Vallor's list of virtues does not seem to correspond with a distinctly new ethical theory, but rather largely accords with Aristotle's theory of virtue. As Mcrae argues, Aristotle's theory of virtue would have been sufficient to provide Vallor with the list of "technomoral" virtues, and it seems insufficiently clear how Buddhist virtue ethics has a crucial influence on this list (Mcrae 2017). Nevertheless, Vallor provides an account of the virtues that is shown to be relevant for technological use contexts in the twenty-first century and that could potentially be embraced by different virtue ethics traditions.

Lack of an Account of Technological Mediation

The shortest way to proceed at this point would be to derive our notion of the ethical aim from Vallor's virtue ethics of technology, and to link this to Ricoeur's theory of technical practice. However, we raise two main criticisms that will lead us to circle back to Ricoeur instead. These criticisms are that (1) Vallor focuses on a one-sided, anthropocentric conception of agency, thereby paying insufficient heed to the notion of technological mediation and that (2) she does not provide a distinct theory of the cultivation of "technomoral" virtue. To support our argument, we will discuss a widely debated case study originating from work in moral psychology around the Milgram experiment that is frequently interpreted as an attack on virtue ethics.

Our first criticism is that Vallor presents a one-sided notion of agency that is too much centred on human agency. Even though she positions her framework as being in line with the "empirical turn" (Vallor 2016, p. 31) of philosophy and ethics of technology (Achterhuis 2001) she does not pay sufficient heed to one of the central concepts of this body of work: namely the concept of technological mediation. On the one hand, Vallor argues that one of the aims of her framework is to offer an account of what *we* can *do* in order to cultivate our technomoral virtues by virtue of which we will be able to cope with technosocial opacity (Vallor 2016, p. 6). This aim, we concede, is attained to a large extent by means of the theory she constructs, notably by stipulating what types of virtues we should cultivate in order to be able to deal with technological change. On the other hand, however, she claims to provide a convincing account of what emerging technologies *do* to *us*, which is a claim that we argue is not reflected in the contents of her theory. This aim resonates with the idea proposed by scholars of technological mediation, foremost perhaps by Verbeek in the notion of *What Things Do* (Verbeek 2005—emphasis added). As such, human virtues, understood as dispositions,[3] as well as the practices by which humans cultivate their virtues, are mediated by technologies. Understanding the process by which technological mediation occurs is therefore crucial for understanding how not only humans can cultivate their virtues, but how humans and technologies can "co-

[3] "Dispositions" and "traits" seem to be handled as synonyms in writings on virtue ethics. However, etymologically, they have distinct meanings. "Trait" originates from the Latin verb "trahere", which means to draw (along) and thereby indicates the position of a drawer with regard to that which is drawn (or: someone projecting a trait on something, saying (on inspection) "this knife has the trait of sharpness"). "Disposition", however, originates from the Latin verb "ponere", which simply means to place, to put (to have a disposition is to tend to place oneself, to put oneself in a certain way) (Partridge 2006). Disposition, therefore, belongs more clearly to the agency of a subject (being disposed to *x*), while a trait belongs more clearly to the knowledge of an observer (knowing something to have a certain trait). As a conclusion, we will refrain from using the term "traits" when discussing the virtues and use the term "dispositions" instead.

cultivate" the virtues. By introducing this less anthropocentric concept we do not, of course, suggest that technologies themselves can be virtuous, but rather that the cultivation of human virtue crucially depends on the agency of technologies and how we relate to that agency, as argued by philosophers of technology in the phenomenological and hermeneutic tradition.

To illustrate the problem of the missing notion of technological mediation, we focus on a point in Vallor's work at which she defends her virtue-based approach against criticisms of virtue ethics coming from moral psychology. She discusses the argument that is related to the famous Milgram experiments in which "research subjects were asked to 'punish' a screaming 'victim' with realistic (but simulated) shocks at the polite request of an experimenter" (Merritt et al. 2010, p. 363). Even though the "victim" was an actor who had to pretend that he was hurt by the simulated electric shocks, the experimental setup was sufficiently convincing for the participants to be conceived as being real. The overall outcome of this experiment was that 33 out of 40 research subjects continued the experiment after initial protests of the victim and that 26 of them continued through agonised screaming and a final unresponsive silence. Moral psychologists argue that experiments like this one show "that the difference between good conduct and bad appears to reside in the *situation* more than in the person" (Merritt et al. 2010, p. 357—emphasis added). This is conceived as an attack on the notion of human character that is central to virtue ethics, because our character allegedly should enable us to act morally right in a consistent manner. One could start to criticise this interpretation of the findings by arguing that the concept of "situation" is black-boxed in this approach and that the particular context of the experiment is left un-examined. However, Vallor takes another approach to defend virtue ethics, arguing that her approach holds because virtue is "*by definition* exemplary rather than typical" (Vallor 2016, p. 22). The fact that a minority of research subjects refrained from following the requests of the experimenter is seen as decisive for acknowledging the correctness of virtue ethics. In other words, the agency of the human subjects who resisted the request for engaging in the experiment is taken as a reflection of a categorical[4] form of their strength of character that in turn *proves* the validity of a virtue ethics approach.

This defence seems to be unsatisfactory for a number of reasons. First, the claim that virtue is cultivated by means of following the practices of exemplary figures does not entail that virtue is "by definition" exemplary rather than typical. Instead, virtue ethics traditions allow for different degrees in virtuous character (hence, having a degree of virtuous character *is* typical), in which exemplary members of a certain moral community (the *phronimos*, in Aristotle) display the highest degrees. Particular communities can consist of only vicious or only virtuous people or can be mixed. Second, different experiments in moral psychology show different levels of "virtuous behaviour" (Merritt et al. 2010, pp. 356–357),[5] understood as consistency in engagement in particular actions, which implies that someone whose virtuous character enables her to cope with one challenging moral setting might be unable to cope with another, depending on the context. This does not mean that such a person's character is not "virtuous", but instead that whether the degree to which her virtuous character is sufficiently cultivated to deal with a particular situation depends in part on the situation, or is *mediated* by the situation, and not merely by the agent's character. If we accept this claim, moral psychologists seem to be at least to some extent warranted in drawing their conclusions. Third, the supposed warrant that a minority of people will be able to deal with situations like the one presented in this experiment, which—importantly—is *not* technosocially opaque (people are aware of the way in which their actions conducted through technologies have certain consequences), is one that no moral philosophy can be satisfied with and especially not one that introduces the additional factor of technosocial opacity. In a world of technosocial opacity, the virtuous conduct of a

[4] Their virtuous character is to be interpreted categorically because, for Vallor's argument of "proof" to hold, the research subjects resisting to engage in the experiment should remain to do so—even in different, more extreme experiments or situations.

[5] Notably, four out of five of the experiments presented by Merritt et al. have a technology as an important mediator of the "situation", be-it money, a prison (as a disciplinary technology), or mechanical noise (Merritt et al. 2010: 357). One example is ambiguous, for the variable is "being in a hurry", but one does not need many persuasive arguments to at least consider the possibility that "being in a hurry" in our contemporary society is highly technologically mediated (e.g., see Bucher et al. 2013).

small minority is not sufficient for safeguarding the flourishing of humanity; and can even less be expected given the greater complexity and opacity as compared to the Milgram experiment. For instance, we can consider whether a non-digital native with the highest degree of virtuous character would be sufficiently capable of dealing with the novel setting of ubiquitous digital technologies. The answer would probably be negative.

This argument does not disqualify the project of constructing a virtue ethics of technology. Rather, it shows that "technomoral" virtues are co-shaped by both people's characters *and* by "settings", that is, by the situation and the environment. This leads us to the role of technology. A philosopher of technology would criticise the conclusions drawn by moral psychologists on the basis that the factor of technological mediation is fully "black-boxed" by sketching the concept of "setting" as something that is somehow separated from a "person". Instead, she would argue that on the one hand "setting"—in contrast to what Vallor argues—should be taken seriously, but that on the other hand the concept of setting should be turned into a concept of technological mediation. Especially in the case of the Milgram experiment, the factor of technological mediation seems important for understanding the "situation." The technology of an electronic system being triggered by a voltage lever constitutes a process of what can be designated as "distancing" (Coeckelbergh 2015): the research subject does not simply "hurt" a victim, but hurts the victim in a mediated way, through a device that presents no feedback with regard to the severity of the harm and in an architecture that separates the research subject from the victim and thereby creates a moral distance between them.

Recent scholarly work that revisited the Milgram experiment seems to be largely in agreement with the above argument. For instance, Haslam et al. (2014) argue that the setup of the Milgram experiment, which included aspects like its "location, the appearance and behaviour of the experimenter, and the *technical apparatus*" (Haslam et al. 2014, p. 275—emphasis added) mediated the interpretation of research participants of their actions as taking place in the context of the paradigm of scientific research. In other words, the situational aspects provided research participants with an interpretative framework (related to the authority of science) through which they could understand the technical practice they were engaging in. In a similar vein, Burger (2014) discusses situational aspects of the experiment that mediated the actions of the research participants. He explicitly focuses on the role of the technical device used in the experiment (a lever indicating voltage levels), arguing that both its detachment from the victim's responses and its capacity to incrementally administer electric shocks contributed to the choices made by the research participants (Burger 2014, p. 492). Even though these scholars do not unpack the notion of technological mediation, they provide good reasons for integrating the notion of "setting", which includes the use of technology, in a virtue ethics account of the Milgram experiment.

The illustration of the Milgram experiment points at a general tendency that runs throughout Vallors work, of emphasising human agency in dealing with technology and insufficiently discussing the impact of technologies on this agency. This can also be observed in the solutions she puts forward for problems posed by technology, be-it "improved techno-moral education" (Vallor 2016, p. 204), "cultivating technomoral humility" (p. 207) or cultivating "renewed technomoral courage" (p. 218). Next to these suggestions, no systematic guidance[6] is provided regarding the way in which we can understand how technology design impacts the virtues and how we could design technologies to enhance the cultivation our virtues. In other words, Vallor offers systematic theoretical guidance for cultivating the virtues to deal with technological change, but fails to offer similar guidance for designing technologies in ways that cultivate and do not obstruct the virtues. Accordingly, we argue that Vallor's virtue ethics of technology needs an account of technological mediation to be suitable as a basis for our novel method.

Our second criticism is that Vallor does not provide a distinct theory of "technomoral" virtue, in the way characterised by MacIntyre (MacIntyre 2007, p. 183), that is, what is missing is an account of virtue related to practice. Vallor concedes that she answers the "what" question of virtue by providing a list of "technomoral" virtues. However, the more

[6] Vallor does occasionally discuss specific design features, for instance concerning data storage for child safety applications (Vallor 2016, p. 206). However, these discussions are largely anecdotal and are not supported by a theoretical, systematic account of how the technology design (e.g., child safety application) mediates the virtues and how different alternatives could either enhance or obstruct cultivation of these virtues.

fundamental "what" question that is posed in virtue ethics, namely "what is virtue" and—by extension—"what is technomoral virtue" remains largely unanswered by Vallor, which is reflected in the varied and loose way in which she defines virtues, which are being referred to as "traits", "dispositions", but also as "discerning skills" and "recognitions" (Vallor 2016, pp. 37, 126). Even though she indicates that her account of virtue ethics is in accord with MacIntyre's, she does not seem to pay sufficient heed to his argument that his theory of practice is what *unifies* the different theories of virtue he discusses (MacIntyre 2007, p. 191) and does not substantively engage with MacIntyre's account of practice. In fact, Vallor takes a different approach, by pragmatically synthesising conditions for cultivating virtues and lists of virtues in different virtue ethics traditions, without however providing a distinct, unified theory of virtue. We argue that even though the synthesising of different virtue traditions has been fruitful, not paying heed to the central role of MacIntyre's theory of practice has led to a missed opportunity of providing a distinct theory of virtue that draws from contemporary reflections on technology and technical practice.

MacIntyre's most notable contribution to the field of virtue ethics has been his theory of practice that largely incorporates and goes beyond Aristotle's conditions for virtuous practice as presented in Book 3 of *Nicomachean Ethics*.[7] That is, like Vallor, Aristotle provided an account of reasoned *conditions* of virtuous practice: that it should be voluntary, based on a rational decision, be open to deliberation, and be in accordance with the wish of the agent; leading to the conclusion that the virtues and vices are within the agent's power. For Aristotle, these conditions ought to be met for a practice to lead to the cultivation of the virtues. What MacIntyre crucially adds to Aristotle's conditions of praxis, is an account of how a practice relates to a human life and to society as a whole (or, as we discussed above, he adds a complementary theory of social reality). Going back to the discussion of Milgram's experiment and the crucial notion of its technologically mediated "setting", we can see how MacIntyre's theory

[7] Even though MacIntyre surpasses Aristotle in this respect, he at the same time acknowledges that his account "captures much, but very far from all, of what the Aristotelian tradition taught about virtue" (MacIntyre 2007, p. 203).

of practice might offer a better starting point than Vallor's synthesised conditions for virtuous practice. That is, MacIntyre enables us to understand the practice of participating in a scientific experiment and the role of symbolic and technological mediations in this practice. For instance, his theory allows us to show that causing harm to a victim as happened in the Milgram experiment relates to a socially established activity through which goods *external* to that activity (honour or monetary reward as result of participating) are realised. He thereby points at the importance of considering the setting of an *institution* in which the experiment takes place that is the social bearer of the practice (MacIntyre 2007, p. 195).

Considering the Milgram experiment in light of MacIntyre's theory, we could state that the institutional setting of participants and the "victim" being in different rooms mediated the practice in question. Participants were not able to directly communicate with the victim and for instance ask the victim questions or touch the victim's body. They only had mediated and limited access to the alleged victim's suffering, comparable to some degree for instance with a drone pilot attacking targets in a desert in Afghanistan. Moreover, the "action at a distance" as enabled by the pushing of a button mediated the practice by preventing the participant from receiving direct feedback, as would for instance be the case when punching the victim and being confronted with a verbal or bodily response. These are significant aspects to consider when discussing the activity as a "practice" that either obstructs or cultivates the virtues of the participants in the experiment. Considered at a higher level, of "narrative structure", the question of "what are you doing" will not merely be answered by "pushing a button to apply electric charges to the victim". Instead, it will be answered in narrative form, which includes accounts of participating in scientific experiments, relating to an authoritative scientist and of the alleged system behind the button and the wires connected to the alleged victim. Finally, we can consider the practice as well as the narratives through which it is rendered intelligible as embedded in a moral community. In this community, science has a certain authoritative status; technology is often considered as a neutral instrument, and so forth.

Having traversed the detour through contemporary work on virtue ethics of technology, where do we now stand? We have seen that Vallor's theory is incomplete for our purposes; for serving as an account of the ethical aim of technical practice. Yet, she does provide us with a comprehensive heuristic of virtues that are both shown to be relevant in the current age of technological change, and potentially resonant at the global level. For this reason, we will integrate Vallor's heuristic of virtues in our account of the ethical aim of technical practice, with, however, some notable adaptations. First, instead of wisdom we refer to the virtue of prudence, in accordance with the Aristotelian concept of phronesis. Phronesis is in the leading translations of Aristotle's work translated as prudence, and wisdom translates into another virtue of thought: of *sophia*. Hence, we align ourselves with the leading translations, in order not to confuse prudence as *phronesis* with the Aristotelian virtue of wisdom, *sophia*. Second, empathy is rephrased as friendship, in accordance with the criticism that empathy itself should *not* be considered a virtue because it is a skill when considered as a voluntary and it is a capacity when considered as an involuntary disposition (Battaly 2011). We therefore retain Aristotle's related virtue of friendship. Third, we sustain the Aristotelian distinction between virtues of character and virtues of thought, which implies that the missing virtues of thought are added (next to prudence: craft knowledge, scientific knowledge, wisdom and understanding).

The Ethical Aim

In this section, we aim to complete our concept of virtuous technical practice to the furthest extent, by returning to Ricoeur's notion of the ethical aim. As stabiliser of the unstable mixture of fabulation and actual experience, the ethical aim points at the *telos* of praxis, the good life for a human being (Aristotle's eudaimonia). This notion completes the determination of virtue to the furthest extent, by restating it as being awake and succeeding for the sake of (1) the good life, (2) with and for others, (3) in just institutions. The ethical aim stipulates a three-fold movement, starting from a naïve conception of phronesis, through its actualisation by means of the norm, to a critical conception of practical reason. It is crucial to understand that by positing this movement of the ethical aim, Ricoeur aims to arrive at an account of practical reason that reconciles the deontological tradition, grounded in Kant's notion of *Moralität* (morality), with the teleological tradition, focusing on Aristotle's notion of *phronesis* and Hegel's notion of *Sittlichkeit* (also usually translated as "morality") (Ricoeur 1992, p. 290).

Ricoeur calls his ethical treatise in *Oneself as Another* his "little ethics", which despite its ironically modest title has an ambitious aim as we can observe above. What the three-fold movement of Ricoeur's ethics offers us is a way in which we can reflect on technical practice by considering how it relates to the ethical aim. Thereby, this ethics provides the final determination of the notion of technical practice that cultivates the virtues, which allows us to distinguish the "virtuous" assassin from the virtuous human being. As such, it takes into account the self (the good life for me), the interpersonal (the good life for others) and the societal (the good life mediated by just institutions), which are interdependent (Ricoeur 1992). It thereby covers the entire scope of technical practices: the ones we perform for our wellbeing, with and for other people, and for contributing to our shared institutions. Each of these three elements of the ethical aim, according to Ricoeur, has to move through the "sieve of the norm" (Ricoeur 1992, p. 170), through their actualisation in a normative structure, and eventually through a reflection on the norm based on a refined conception of the ethical aim. In what follows, we will trace this movement according to the three stages of the ethical aim: (1) the good life, (2) with and for others, (3) in just institutions. For each stage, we highlight the way in which it further determines the notion of "technical practice that cultivates the virtues".

The Good Life

Ricoeur's conception of the ethical aim seems almost diametrically opposed to Heidegger's notion of being-towards-death (Heidegger 1996), because it does not attempt to ground itself in an existential analysis but instead traverses aspects of life from the self, extending all the way towards the other and public life. For Ricoeur, the "naïve", or perhaps rather "pre-dialogic", initial conception of the good life that makes it possible for people to engage in technical practice virtuously, is founded on the notion of "self-esteem", mediated through "self-interpretation" (Ricoeur 1992, p. 179). This conception is directly tied with the threefold structure of practice, that moves through constitutive rules[8] and life plans to the narrative unity of life. As such, we can evaluate our actions in the narrative mode and can accordingly evaluate ourselves as the authors of these actions. For instance, we can use the standards of excellence that govern the practice of "playing the piano" to evaluate ourselves as the author of this practice (e.g., the "who" performing the musical piece), which we can extend in light of a life plan (e.g., the "who" that wants to become a professional piano player) and in light of the narrative unity of life (e.g., the "who" that might choose between being a piano player, a teacher, and a soldier). When it comes to the narrative unity of life, we face an open horizon of more or less distant ideals that asks for a never-ending process of interpretation (Ricoeur 1992, p. 180). This open-endedness points at human beings understanding themselves as both acting *and* suffering beings. That is, in making life-choices, for instance choosing whether to become a professional pianist or a soldier, we only ever *co-author* the narrative structure that makes such choices intelligible. Unforeseen events might come in between and make us "suffer", for instance if a physical test shows one to have insufficient eyesight to join the army and thereby closes off that choice. What this tells us about technical practice, is that the minimum requirement for such a practice to be engaged in virtuously is that a practitioner can interpret her self-esteem not only through evaluating the respective practice (e.g., creating a good computer programme), but also through the related life plan(s) (e.g., wanting to *be* a good programmer) and narrative unity of life (e.g., having chosen well to become a programmer vis-à-vis numerable other choices). This shows us for instance that it is both extremely difficult and undesirable to aim to have someone engage in "ethical practice" concerning a technical practice that he cannot properly evaluate (e.g., a

[8] Constitutive rules, Ricoeur argues, as standards of excellence, provide practices with the notion of goods that are integral to them, and can when turned into norms provide contents to the categorical imperative (Ricoeur 1992, p. 176).

"mindless" task), related to a life plan that he feels alienated from (e.g., a pointless job), and a unity of life that he cannot make sense of (e.g., having no clue of why this job was chosen). This might seem to be an issue that commonly the human resources department of an R&I institution deals with, but Ricoeur shows how it is also firmly tied to the ethical aim as such.

The pre-dialogic notion of the good life as self-esteem needs to be actualised, which makes Ricoeur turn to Kantian moral philosophy and the norm. With the notion of the "sieve of the norm" (Ricoeur 1992, p. 215), Ricoeur accomplishes an original cross-fertilisation between Aristotelian and Kantian ethics. That is, he argues both for (1) the necessity of the ethical aim to move through the deontological critique and for (2) the ultimate grounding of the deontological viewpoint in the teleological perspective. Translating the naïve ethical aim into a normative structure is necessary, according to Ricoeur, for two reasons. First, the facticity of an inclination towards evil in the world makes it insufficient to merely have a notion of the good life, but demands the formulation of moral laws that traverse the opposite direction, accruing to "thou shall not be evil" (Ricoeur 1992, p. 218). Second, in exercising freedom, and in being able to make free choices, we find the fact that we are always free to choose to follow either good or evil maxims. Thus, we need normative structures ensuring that we act in accordance with the maxim that *limits* our actions so that it promotes the wish of what we ought *not* to be, namely: evil. Deontology therefore provides the necessary limitations on praxis in terms of normative structures, which could not be provided by the naïve conception of the good life. In Kantian ethics, this idea is captured by the notion of *autonomy*, which implies obedience to oneself, or self-legislation. Ricoeur connects this idea of self-legislation with constitutive rules that determine the standards of excellence of a practice, namely by stating that the categorical imperative is of the same form: that it internalises a constitutive rule that posits limits on action and at the same time provides a ground for appraisal. By means of self-legislation, we progress from self-esteem to "self-respect", which is "self-esteem under the reign of law" (Ricoeur 1992, p. 215). What this tells us about technical practice that cultivates the virtues, is that the practitioner ought not only to be able to evaluate her practices, life plans, and narrative unity of life, but also to limit her choices regarding these practices based on moral laws. This sense of limitation might be accomplished through performing an ethical oath, for instance in professional settings in banking and finance as advocated by Blok (2013). Blok argues that an ethical oath can mediate a technical practice when it takes place in a public context and when it pertains to the sense of self of the practitioner taking the oath. This accords with Ricoeur's theses that (1) selfhood is brought about by "keeping one's word" (Ricoeur 1992, p. 123), and (2) that the sense of obligation derived from Kantian ethics does not imply any moral solipsism, but rather a moment of universalization that involves "neither you nor me" (Ricoeur 1992, p. 204), which therefore essentially relates it to public life. In other words, by means of publicly establishing the moral limits of a technical practice, an practitioner can provide content to the categorical imperative as a constitutive rule that mediates her self-respect in acting virtuously.

Finally, Ricoeur shows how self-esteem, through self-respect, turns into *conviction* by reflecting on the application of the norm. By means of this final step, Ricoeur shows that engaging in a technical practice while following a norm is *not enough*, and that one always needs to reflect on this norm by looking at the context in which it is applied and keeping the ethical aim in mind. This claim results from a questioning of the universalism of Kantian ethics from the contextualist viewpoint. Most notably,[9] Ricoeur focuses on the encounter between the plea for universality of

[9] Ricoeur provides two additional reasons for questioning Kant's universalism. First, he argues that the Kantian notion of autonomy is really a *political* term that has been turned into a moral one, which depends on the heteronomy of public life rather than on the universalised individual. This is the case because autonomy presupposes the reciprocity of justice and the respect for other persons, in assuming that the self should be treated equally to the other who deserves respect and to the most disadvantaged in society who should be the focus of distributive justice. Consequently, in order to become truly "autonomous", we need the guidance of the *phronimos*, of the master of justice who teaches us how to take responsibility for our own choices (Ricoeur 1992, p. 276). Second, Ricoeur criticises Kant's criterion of universalization, by stating that the coherence of a system of norms that implies a plurality of duties cannot be derived from the non-contradiction of duties with the categorical imperative. Instead, he states that a system of norms gains its coherence from being *constructed*, by way of refining norms contextually such as in the system of common law. For instance, the context provided by the case of "self-defence" allows for the refinement of the norm "thou shall not kill", by adding a specifying premise. At the same time, such a construction of a normative system ought always to be submitted to a criticism of the potential prejudices and ideological assumptions that might underlie the choices for specifications.

norms and the weight of the historicity of concrete morality (Ricoeur 1992, p. 280). When universalist norms are put into practice, they can instigate conflicts based on the historical and cultural context of their application. For instance, Article 17 in the Universal Declaration of Human Rights on the right to property might cause conflict in societies that offer no cultural and historical basis for a compatible conception of personal property (Donnelly 1984). The more specific certain norms become, the more likely they are to instigate such conflicts, and Ricoeur sees the essential diversity of goods that plays a role in such conflicts—and consequently the impossibility to have a *single* conception of distributive justice—as the reason to support the contextualist objection to universalism. For contextualists such as MacIntyre, the justification for norms as standards of excellence is traced back to a historical community, whereas for universalists such as Habermas this justification lies in a-historical principles of communicative action. Ricoeur situates himself at the crossroads between these traditions, by accepting Habermas' project of creating an "ethics of argumentation" (discourse ethics) while also renouncing the idea of ultimate foundation of this ethics in a-historical principles and consequently acknowledging the need to follow inverse path from that of justification: namely that of the actualisation of a normative system through practices, life plans, and the narrative unity of life.

Hence, Ricoeur states that ultimately discussions about autonomy, self-respect and the rule of justice should be mediated by a *reflective equilibrium* between the ethics of argumentation and "considered convictions" (Ricoeur 1992, p. 289). He argues that argumentation, as an institutionalised language game, always involves other language games, including the crucial one of "narration", which grounds arguments in the world of action that is historically and culturally situated (Ricoeur 1992, p. 288). The "best argument" concerning a discussion about norms is therefore one that not only fulfils formal principles of communicative action, but that also finds its contextual support in what MacIntyre calls a historical "moral community". To arrive at a reflective equilibrium between the ethics of argumentation and considered conviction, Ricoeur argues that we have to assume the paradox that "one must maintain the universal claim attached to a few values where the universal and the historical intersect" *and* that "one must submit this claim to discussion, *not* on the formal level, but on the level of the convictions incorporated in concrete forms of life" (Ricoeur 1992, p. 289). The term "value" here is centrally used for the first time by Ricoeur and designates a "compromise concept" rather than a genuine moral or ethical concept. That is, a value can only arise from compromises made between universalist arguments and historical convictions. What this means for technical practice that cultivates the virtues is that the normative structure that governs a certain practice ought to rely *both* on a few values that are universally recognised *and* on the critical discussion of these values within a moral community. For instance, the building of medical equipment that relies on the use of embryos should be governed by universally recognised values, such as those presented in the UNESCO report on the use of stem cells in therapeutic research (Smith and Revel 2001), which are justified by considering the universalization principles in discourse ethics. At the same time, these values should be submitted to discussions about the context of their application, for instance, about the ambiguous difference between "therapy" and "enhancement". Following Ricoeur, such a difference cannot be established by means of applying a-historical principles but can only arise from the use of narratives and other language games that provide the necessary context for arguments that try to establish the difference. Consider how a narrated account of an innovation relying on embryonic research, one that makes explicit how this innovation enabled a disabled child to suddenly be able to communicate with other children and partake in everyday life, would elucidate how this innovation relates to the "therapy-enhancement" distinction.

With and for Others

For Ricoeur, the notion of the good life already depends on a dialogical structure that includes otherness, but this otherness is only made explicit by the addition of "with and for others". Again, Ricoeur differs from Heidegger by making *care* dependent on the life with and for others, reconceptualising it as *solicitude*. That is, we can have self-esteem only to the extent that we care for the other as oneself (you appear before me as an "I") and for oneself as another (I evaluate myself according to the *phronimos*, the esteem of others). Departing from the Aristotelian virtue of *friendship* as an expression of this mediation, Ricoeur states that friendship for the sake of the good needs to be distinguished from friendship for the sake of utility and pleasure—for the latter two are essentially self-regarding. Friendship for the sake of the good is to be based on a mutual relationship between oneself and another, on a basic sense of reciprocity that involves a relation of *giving* and *receiving* (Ricoeur 1992, p. 188). On the one hand, when the other takes the initiative we receive from the other what is lacking in ourselves by gaining from another's instruction and we give to the other our esteem, which leads to mutual recognition. On the other hand, when we take the initiative, we give our sympathy and care to the other who is *suffering*, while at the same time receiving a sense of vulnerability related to the condition of mortality. Hence, this relation implies the mutual constitution of "agent" and "patient" in the process of giving and receiving. In a technical practice that cultivates the virtues, agents and patients are caught up in a reciprocal relationship that joins together "the reversibility of roles" (you regard me as an "I" and vice-versa) and the "non-substitutability" of persons ("each person is irreplaceable in our affection and esteem"), which we spontaneously assume because of the similitude between oneself and another (Ricoeur 1992, p. 193). In other words, when engaging virtuously in a technical practice, we spontaneously assume that for-the-sake-of-which the practice is, is not merely related to the good life for ourselves but also for others as acting and suffering, who are irreplaceable and essentially like ourselves. For instance, when we engage in the technical practice of "conversing on Facebook", it is not sufficient to link the for-the-sake-of-which of this practice to our personal idea of friendship in the good life, which risks being limited to self-love. Complementarily, we should regard *others* implied in this technical practice, by considering whether it allows us to attend to their suffering and receive their instructions. When electronic messaging on Facebook is merely performed to receive the esteem of others, no reciprocal relation is established, and no solicitude is implied that is necessary for the ethical aim.

As with the pre-dialogic notion of the good life, our understanding of solicitude needs to move through the sieve of the norm in order to be actualised as a "respect for persons". For this, Ricoeur turns to the Jewish and Christian conceptions of the "golden rule", which can be stated negatively, in saying that one should *not* do to others what one would hate to do to oneself and positively, in saying that one should *love* another as oneself (Ricoeur 1992, p. 219). The positive formulation, departing from solicitude, focuses on agency and includes the possibility of exerting power over someone else, which culminates in the most problematic cases in humiliation: the destruction of another's self-respect. The negative formulation, departing from reciprocity, focuses on patiency (i.e., being a patient) and balances the positive formulation by placing a limit on practice (thou shall not lie, not kill, not humiliate, and so forth). Ricoeur argues that Kant's second formulation of the categorical imperative is a formalisation of the golden rule, because it superimposes the universal concept of humanity—that by reason of which each is made worthy of respect—on the diversity implied by agency and patiency (i.e., always treat *humanity* never simply as a means, but always at the same time as an end). However, Ricoeur also shows that Kant cannot bring this formulation to its logical conclusion because it still depends on the essential notion of respect of persons, which can only be demonstrated by grounding the "fact of morality" on the fact that persons exist and that the end of a human being is distinct from the end of things that have utility and can bear a price. Ricoeur argues that this fact of morality can only be explained by pointing back at solicitude, at the care for another as oneself. For Ricoeur, therefore, respect for persons means restoring reciprocity where there is a lack of it (Ricoeur 1992, p. 225), on the basis of a plurality of persons existing as ends in themselves. What this tells us about virtuous technical practice, is that everyone consciously affected by such a practice—both acting and suffering—ought always to be considered as a person whose existence is an end in itself and that a balancing of agency and patiency ought to be aimed at. For instance, with regards to practices of automation in innovation settings, the workers being replaced should not be regarded as mere means (replaceable human resources), but always at the same time as ends in themselves (employees who deserve respect), whose patiency should be compensated by allowing them to have a saying in the way the automation process takes place and the way in which reciprocity (e.g., compensation arrangements) is ensured.

Finally, Ricoeur shows that following the norm of treating other persons with respect in a technical practice is not *enough*, and that one needs to reflect on this norm, which allows Ricoeur to arrive at "critical solicitude" (Ricoeur 1992, p. 273). He argues that conflicts arise when the "otherness of persons, inherent in the very idea of human plurality" (Ricoeur 1992, p. 262) in certain contexts proves to clash with "the universality of rules that underlie the idea of humanity" (ibid.). In order to question Kant's notion of humanity, Ricoeur considers the possibility of making an exception on a universal rule based on the otherness of another person. For this, he considers the important case of "keeping a promise" where a constitutive rule is invoked stating that one places oneself under the obligation of doing such and such in specific circumstances. Ricoeur argues that this constitutive rule only confronts us with a moral problem on the basis of the *reason* for keeping the promise, which he posits as the principle of fidelity. Such a moral problem involves a dyadic structure, meaning that it involves two persons who both relate to the reason for keeping the promise, which implies that next to my will to keep a promise there is another person's expectation(s). Therefore, "it is to the other that I wish to be faithful" (Ricoeur 1992, p. 268). This shows a distinction between legal promises such as contracts, which require something of someone regardless of the other, and keeping a promise, which implies someone *counting* on someone else, expecting him to keep his promise.

In light of this distinction, Ricoeur shows how exceptional circumstances can cause one to consider either to act or to refrain from acting in accordance with a moral rule based on the respect for persons, which essentially derives from solicitude. For instance, he considers the case of telling the truth to the dying. In case the dying person is morally or physically too weak to hear the truth, one should be compassionate and consider refraining from telling the truth. Similarly, questions arise when the distinction between "thing" and "person" to whom one is obliged to act in a certain way is ambiguous, like in the case of embryos conceived in test tubes for scientific research. This is one of the few places in his work where Ricoeur explicitly refers to modern technology, by stating that certain technical practices such as ones involving the use of embryos in innovation settings render the dichotomy between things and persons ambiguous and therefore problematic. He also refers to the case of the impacts of emerging technologies on future generations, whose personhood is not firmly established, as another context that problematizes the universal application of rules connected to the respect to persons. To tackle these problematic cases, Ricoeur posits the notion of *critical solicitude* as the refined version of respect for persons, and adhering to three features of practical reason: (1) the recognition that a conflict between a rule and its application involves different positions adhering to the same principle of respect, (2) the positioning of Aristotle's notion of the mean as guiding principles, while accepting that this might imply enforcing a dichotomy, and (3) to take guidance of the *phronimos* into account. What this tells us about technical practice, is that in cases when such a practice involves problematic contexts of personhood—notably, (1) through lack of cognitive or emotional capacities, (2) through lack of a distinction between personhood and "thinghood", and (3) through the fog of time, considering future persons—the norms governing the practice with regards to respect for persons ought to be receptive to adverse positions, seek the mean, and involve council of the *phronimos*.

In Just Institutions

To complete the determination of the ethical aim, Ricoeur moves from the interpersonal, characterised by solicitude, to the *institutional*, characterised by justice. Ricoeur thereby notably distances himself again from MacIntyre, by repudiating MacIntyre's claim that all institutions are necessarily concerned with external goods and thereby diametrically opposed to practices (MacIntyre 2007, p. 194). Instead, Ricoeur acknowledges the intimate tie between practices and institutions, showing how institutions mediate conflicts that can arise from political practices. Starting again at the pre-dialogic stage, he argues that an institution denotes "the structure of *living together* as this belongs to a historical community", which secures duration, cohesion, and distribution (Ricoeur 1992, p. 194). He derives this initial ethical understanding of institution from Arendt's concept of "power in common" that is contrasted with domination ("power over") and that is realised by a plurality of people acting in concert. The notion of plurality extends the face-to-face encounter of the interpersonal relation to acting with third parties that are not directly known. Because acting in concert with unknown third parties needs time to unfold, institutions are needed. Institutions thereby provide the necessary temporal dimension for the power in common to endure, which lies at the basis of any political community. This understanding of institutions is bound up with the Aristotelian virtue of justice, which Ricoeur argues does not yet belong to the deontological notion of procedural justice. Initially, justice—like friendship—is aimed at the good, which by extending the interpersonal towards unknown third parties becomes the common good. For this extending to unfold, people do not only partake in institutions, but also *distribute* the related "parts" (roles, tasks, effort) within the political community. The ethical sense of justice thereby denotes a demand for equality, which is linked to solicitude at the interpersonal level by extending the idea of "the same for you as for me" to "the same for each", which is achieved through distribution. What this tells us about technical practice that cultivates the virtues is that the for-the-sake-of-which of such a practice ought not only to regard the known other, but also the unknown other in a political community. For instance, the technical practice of "making an investment" not only needs to regard the known parties directly involved (making a "fair" deal for the seller and potential buyers), but also needs to *refer back* to the world of action—as established in the narrative technologies approach—and consider a notion of fairness for those affected by the technical practice who are unknown: for instance the families affected by the investment practices into toxic mortgages prior to the 2008 financial crisis.

Ricoeur argues that *the just* faces in two directions: towards the good, connected with the pre-dialogic, ethical sense of justice, and towards the legal, connected with the moral *rule of justice* (Ricoeur 1992, p. 197). Our sense of justice needs a normative structure to clarify its inherent ambiguities: (1) concerning the foundation of a "just share" on the basis of separation or cooperation, (2) concerning the justification for distinguishing between arithmetic equality and proportional equality, and (3) concerning the tension between justice based on the extension of solicitude and based on a system of impersonal positive law. In other words, even though we have an initial sense of justice based on the extension of solicitude, we need to actualise this sense of justice through a normative structure of rules, which demands justification. In order to justify a particular normative system, the fiction of the social contract is posited to replace the common good with an imaginary procedural deliberation (Ricoeur 1992, p. 228). Ricoeur argues that the social contract has a function similar to Kant's notion of autonomy but differs from autonomy because it is not a "fact of reason", but an imaginary political reality. He sees Rawls' conception of justice as fairness as the furthest development of this thought, founded on the a-historical fiction of the veil of ignorance as the original position, which is the basis for a procedure towards a shared agreement on the principles concerning what is just. For instance, it stipulates when arithmetic equality should be applied, based on the claim to basic needs (Rawls 1971, p. 286).

Ricoeur accepts Rawls' justification of the two principles of justice[10] based on the decision on distribution of goods made in a society behind the a-historical veil of ignorance as a formalisation of the sense of justice. However, he rejects this justification as the final determination of justice because it presupposes an understanding of the just and unjust that is still founded on an ethical understanding of justice. In other words, for participants behind the veil of justice to distinguish a just from an unjust distribution of goods based on a procedure, they still have to appeal to a notion of the good, which opens up the problem of a diversity of goods and of incommensurable but equally valid arrangements of distributive justice. Despite this reservation, the normative structure introduced by the social contract adds to the ethical aim and informs the notion of technical practice that cultivates the virtues. That is, the sense of justice implied in the for-the-sake-of-which of a technical practice needs to be submitted to a normative test involving the principles of justice. Considering the example of "making an investment", we could state that we cannot simply extend our sense of justice from our relationship with buyers and sellers to the unknown people affected by our investments.

[10] These two principles are: (1) "each person is to have an equal right to the most extensive total system of equal basic liberties compatible with a similar system of liberty for all" (Rawls 1971, p. 220) and (2) "each person is to have an equal right to the most extensive scheme of equal basic liberties compatible with a similar scheme of liberties for others", while "social and economic inequalities are to be arranged so that they are both (a) reasonably expected to be to everyone's advantage, and (b) attached to positions and offices open to all" (Rawls 1971, p. 53).

Even if it would be "just" for a seller to lose a great deal of profit because of a transaction that was nonetheless conducted fairly, the same would not apply when the same sense of justice would imply that a family would need to vacate its house and risk its livelihood. Therefore, Rawls' principles of justice should be applied to test whether the sense of justice can properly be extended or whether another notion of distributive justice is called for. For instance, in this case the particular family might set the benchmark for what is a fair arrangement such that "social and economic inequalities are to be arranged so that they are" … "reasonably expected to be in everyone's advantage" (Rawls 1971, p. 53).

Finally, Ricoeur states that it is *not enough* to submit one's sense of justice to a normative system, but that one again needs to reflect on this system, which brings us from an initial sense of equality, through the rule of justice, to a *sense of justice*. This need springs from the inevitable conflicts that arise when uniform moral principles are applied to complex social reality. Ricoeur initially refers to Walzer's critique of Rawls, which is focused on the problem of the diversity of goods and calls for the conceptualisation of "spheres of justice" whose goods are incommensurable. Ricoeur shows how this critique introduces the new problem of arbitration between the spheres of justice (e.g., on what basis can we demarcate and limit the competition between the sphere of "money and merchandise" and "security and wellbeing"?). This leads him to consider Hegel's notion of *Sittlichkeit*, which denotes concrete morality in a political community culminating in the state as the final arbiter between the spheres of justice. Ricoeur retains from this notion of *Sittlichkeit* its determination as a "system of collective agencies" and "the gradual triumph of the organic bond between men and women over the exteriority of the juridical relation" (Ricoeur 1992, p. 255). In other words, *Sittlichkeit* does *not* denote a system of individual rights and duties (such as Rawls' notion of distributive justice) but a system of arbitration between competing claims in society that by juxtaposing the application of the law with the ethical significance of particular situations moves in a continual progression towards a system of rights that is maximally non-conflictual. Importantly, Ricoeur argues that *Sittlichkeit* can only be realised on condition that the tie between institutions and the spirit of the people is not perverted: that institutions do not posit aims apart from those granted to them by the people. Perversion happens because the initial agency of the "power in common", which relates to people sharing the ethical aim, is *forgotten* in the process of institutionalisation. We argue that with *Sittlichkeit*, we arrive at the "technical practice" of politics, at that with which the political philosopher as a craftsman is concerned.

Ricoeur indeed defines "the political" as "the set of organised practices relating to the distribution of political power, better termed domination" (Ricoeur 1992, p. 257). As such political practices are aimed at addressing conflicts vertically, between the governing and the governed, and horizontally, between competing groups in civil society. Ricoeur offers three ways in which political practices mediate conflicts, which inform our notion of technical practice that cultivates the virtues. First, he argues that the realm of everyday discussions about the order and priority of different spheres of justice ought to be open and negotiable and should not be dominated by scientific or dogmatic arguments (Ricoeur 1992, p. 258). Second, he argues that these everyday discussions should be embedded in a longer-term discussion on the ends of good government, concerning the meanings of political concepts such as "liberty" and "security" and their anchoring in a good constitution that is open to historical re-interpretation. Third, he argues that both aforementioned types of discussions ought to be embedded in a discussion on the process of legitimation (Ricoeur 1992, p. 260). Democracy, which denotes this process, needs to confess indeterminacy concerning the basis of power, law, and knowledge, and commit to a plural, public debate.

To balance democracy, finally, equity—rephrased as the *sense of justice*—remedies the rule of justice where the legislator fails us through over-simplicity. What this tells about virtuous technical practice, is that when such a practice instigates conflicts between spheres of justice or between domination and legitimation, practices towards solving these conflicts should be (1) organised openly and subjected to negotiations, (2) based on a shared conception of good governance, (3) both of which are legitimised by a democratic process and balanced by a sense of justice. A clear example in this regard are conflicts instigated by the practice of obtaining digital evidence in criminal cases, which particularly came to light when the FBI attempted to get access to an iPhone while Apple refused to offer a "back door" into the software of the device (Hack

2016). Such a conflict both involved spheres of justice (Apple belonging to "money and merchandise" and the FBI to "security and wellbeing") and a conflict between domination and legitimation (the FBI having "power over" the data of US citizens, without clearly having its aim aligned with the "spirit of the people"). In this particular case, Apple refused to assist the FBI and the FBI eventually managed to gain access to the device by its own means. However, when applying Ricoeur's principles to this particular technical practice, neither of the parties involved should have been left to their own judgment. Instead, because of the conflicts created, (1) an open public discussion should have been organised between ordinary citizens, that should have (2) either (a) been informed by existing norms for good governance or (b) led to novel and refined norms, and (3) that should have led to a process of democratic legitimisation, which would have provided the necessary arbitration between the parties involved as well as to a balancing of this process by a sense of justice (commonly vested in the judiciary branch).

Conclusion

At this point, we have a comprehensive account of virtuous technical practice. That is, to engage in a technical practice that cultivates the virtues involves being awake and succeeding, which depends on the nesting of finalities, of the in-order-to with regards to the for-the-sake-of-which. This nesting of finalities consists in linking basic actions, action chains and practices through life plans to more or less distant ideals. Life plans, which introduce the process of technological configuration, denote the two-fold movement of ascending complexification and descending specification in the process of technological emplotment. To understand these movements, we should consider the textuality of a technology, the literacy of the respective practitioners, and the temporality and distancing of the process of technological configuration. We thus have a set of powerful concepts that allow us to interpret any particular technical practice to a certain extent. That is, they allow us to interpret a technical practice by providing a descriptive and narrative account of it and by providing the first determination of virtue, namely of being awake and succeeding.

To relate the notion of being awake and succeeding to the necessary notion of the for-the-sake-of-which in living together in a political community, we needed to link technical practices to the ethical aim. By doing so, we argued that we could only speak of a technical practice that cultivates the virtues when the for-the-sake-of-which of this practice accords with the good life, with and for others, in just institutions (Table 5.1):

Reflecting on our critique of MacIntyre in Chap. 2, we argue that adding the ethical aim(s) avoids the relativism that seems to follow from MacIntyre's communitarian framework. The cultivation of virtue in and through technical practices should not only be linked to the ways of doing and narratives within a particular community, but should also be bound to ethical aims and political principles.

Finally, we need to make two important reservations regarding the completeness of our framework. First, just as was the case with phenomenology, narrative contributes to ethics by recognising its *limit* (Ricoeur 1992, p. 245), thereby recognising it as being inevitably incomplete. According to Ricoeur, particularly tragic narrative is not to be considered a source of moral teaching, but as something that opens the path to moral conviction by meditating on the inevitability of conflict in life. In other words, narrative allows us to recognise the inevitable ethical conflicts in our lives without offering an "ethical" solution to those conflicts. Even though our account of virtuous technical practice approximates the good life, with and for others, in just institutions, this good life is still affected by the inevitability of tragedy. We can design technologies responsibly, use them with care for others, and derive our self-esteem from our engagement in technical practice that cultivates the virtues, but this does not salvage us from hubris, from the inevitable accident, disaster, or tragic course of events. Second, we did not and will not engage with Ricoeur's

Table 5.1 The nine core concepts that capture Ricoeur's movement of the ethical aim, which complete the determination of technical practice that cultivates the virtues to the furthest extent

	Ethical aim$_1$	Sieve of the norm	Ethical aim$_2$
The good life	Self-esteem	Self-respect	Conviction
With and for others	Solicitude	Respect for persons	Critical solicitude
In just institutions	Equality	Rule of justice	Sense of justice

speculative attempt to provide his ethics with ontological foundations in the tenth study of *Oneself as Another*. This omission has two reasons: (1) it would lead us beyond our current purposes of providing an ethics of technology, and (2) it would call for an elaborate commentary when put in light of the accounts of technological mediation, which would lead to a distinct research project in its own right. This has as a ramification that the "otherness" implied in our account (i.e., self-esteem in relation to the esteem of others, solicitude as the care for others, justice as solicitude for the unknown other) remains without a firm ontological foundation and has to be assumed without being adequately justified.

In the next chapter, we will take our concept of virtuous technical practice into the reality of actual innovation practices. Focusing on a particular technology (automated border systems) and related technical practices, we unfold a practically oriented method for doing technology ethics. This method consists of three general phases. The first leads us from the description of narrative structures to their interpretation according to our narrative account of technological mediation. The second guides us from the interpretation of technical practices to their evaluation according to standards of excellence, life plans, and the narrative unity of life. Finally, the third phase leads us from evaluation to prescription; scrutinising technical practices according to the way they are conducive of the good life, with and for others, in just institutions.

References

Achterhuis, H. (2001). *American Philosophy of Technology: The Empirical Turn*. Bloomington, IN: Indiana University Press.

Anscombe, G. E. M. (1958). Modern Moral Philosophy. *Philosophy, 33*(124).

Aristotle. (1999). *Nicomachean Ethics* (T. Irwin, Ed.). Indianapolis: Hackett Publishing Company.

Battaly, H. (2011). Is Empathy a Virtue? In A. Coplan & P. Goldie (Eds.), *Empathy: Philosophical and Psychological Perspectives*. Oxford: Oxford University Press.

Blok, V. (2013). The Power of Speech Acts: Reflections on a Performative Concept of Ethical Oaths in Economics and Business. *Review of Social Economy, 71*(2), 187–208. https://doi.org/10.1080/00346764.2013.799965.

Bucher, E., Fieseler, C., & Suphan, A. (2013). The Stress Potential of Social Media in the Workplace. *Information, Communication & Society, 16*(10), 1639–1667. https://doi.org/10.1080/1369118X.2012.710245.

Burger, J. M. (2014). Situational features in Milgram's experiment that kept his participants shocking. *Journal of Social Issues, 70*(3), 489–500.

Coeckelbergh, M. (2015). *Money Machines*. Farnham/Burlington, VT: Ashgate.

Curzer, H. J. (2017). Yesterday's Virtue Ethicists Meet Tomorrow's High Tech: A Critical Response to Technology and the Virtues by Shannon Vallor. *Philosophy and Technology, 31*(2), 283–292. https://doi.org/10.1007/s13347-017-0269-z.

Donnelly, J. (1984). Cultural Relativism and Universal Human Rights. *Human Rights Quarterly, 6*(4), 400–419. https://doi.org/10.2307/762182.

Foot, P. (1978). *Virtues and Vices and Other Essays in Moral Philosophy*. Berkeley: University of California Press.

Hack, M. (2016). The Implications of Apple's Battle with the FBI. *Network Security, 2016*(7), 8–10. https://doi.org/10.1016/S1353-4858(16)30068-X.

Haslam, S. A., Reicher, S. D., & Birney, M. E. (2014). Nothing by Mere Authority: Evidence that in an Experimental Analogue of the Milgram Paradigm Participants are Motivated not by Orders but by Appeals to Science. *Journal of Social Issues, 70*(3), 473–488. https://doi.org/10.1111/josi.12072.

Heidegger, M. (1996). *Being and Time* (J. Stambaugh, Ed.). New York: State University of New York Press. https://doi.org/10.1017/CBO9781107415324.004.

Hursthouse, R. (1999). *On Virtue Ethics*. Oxford: Oxford University Press.

MacIntyre, A. (2007). *After Virtue: A Study in Moral Theory* (3rd ed.). Notre Dame, Indiana: University of Notre Dame Press. https://doi.org/10.1017/CBO9781107415324.004.

McDowell, J. (1979). Virtue and Reason. *The Monist, 62*(3), 331–350.

Mcrae, E. (2017). Finding a Place for Buddhismin the Ethics of the Future: Comments on Shannon Vallor's Technology and the Virtues: A Philosophical Guide to a Future Worth Wanting. *Philosophy and Technology*, Advance Online. https://doi.org/10.1007/s13347-017-0287-x.

Merritt, M. W., Doris, J. M., & Harman, G. (2010). Chapter 11: Character. In J. M. Doris (Ed.), *The Moral Psychology Handbook* (pp. 1–504). Oxford:

Oxford University Press. https://doi.org/10.1093/acprof:oso/9780199582143.001.0001.
Nussbaum, M. C. (1999). Virtue Ethics: a Misleading Category? *The Journal of Ethics, 3*(3), 163–201.
Partridge, E. (2006). *Origins: A Short Etymological Dictionary of Modern English* (4th ed.). London: Routledge.
Rawls, J. (1971). *A Theory of Justice*. Harvard University Press.
Ricoeur, P. (1992). *Oneself as Another* (K. Blamey, Ed.). Chicago: University of Chicago Press.
Smith, A. M., & Revel, M. (2001). *The Use of Embryonic Stem Cells in Therapeutic Research* (No. BIO-7/00/GT-1/2 (Rev. 3)). Paris.
Vallor, S. (2016). *Technology and the Virtues: A Philosophical Guide to a Future Worth Wanting*. Oxford: Oxford University Press.
Verbeek, P.-P. (2005). *What Things Do: Philosophical Reflections on Technology, Agency, and Design*. Pennsylvania: Pennsylvania University Press.

6

A Method for Technology Ethics

Human existence is saturated with a huge variety of technical practices, practices that play a role in family life, professional life, political life, and so forth. However, the *site* at which most new technical practices are inaugurated is rather more specific: namely the site of "research and innovation" (R&I). For instance, computer science labs in universities build new applications in robotics and modes of transport (think of the Hyperloop). Similarly, large global corporations have R&I divisions that seek breakthrough applications for artificial intelligence (AI), facial recognition, and biomedicine. For this reason, the field of technology ethics has been primarily focused on this context of research and innovation. Like other fields of applied ethics, such as biomedical ethics or environmental ethics, technology ethics is therefore not only preoccupied with developing a normative theory, but also with making such a theory responsive to the concrete practical setting of R&I. Having developed a normative theory based on the notion of virtuous technical practice, we now make this final step by relating this theory to the site of R&I.

We do so by constructing a method for practising ethics in R&I. This method is placed in a wider academic and professional discourse of technology ethics and responsible innovation. This discourse has developed a

W. Reijers, M. Coeckelbergh, *Narrative and Technology Ethics*,
https://doi.org/10.1007/978-3-030-60272-7_6

wide range of methods, which focus on different aspects of the R&I process. More particularly, they focus on (1) anticipating potential ethical impacts of R&I, (2) embedding ethics in the R&I design process, and (3) evaluating the outcomes of R&I (i.e., the concrete impacts of technologies) on society (Reijers et al. 2017). Examples of methods in this context are "ethical impact assessment", "ethical technology assessment", and "value sensitive design". What these methods have largely done is make normative ethics responsive and thereby relevant to R&I settings, including practices of design, exploitation, and governance.

However, existing methods show some important shortcomings. First, they have to some extent divorced themselves from ongoing discourses in normative ethics; innovating therefore more in the sense of application than ethics. Second, they largely inherit the gaps in contemporary philosophy of technology as discussed in Chap. 2; namely, a neglect of language, temporality, and the social. Third, and more specifically, they show some more practical shortcomings: (1) they insufficiently deal with the problem of *uncertainty* in R&I, assuming that future impacts are predictable in principle; (2) they mobilise notions of value or virtue that are inadequately grounded in normative theory; and (3) they offer principled approaches for governance of R&I that insufficiently account for how principles are taken up by practitioners (Reijers et al. 2017). Furthermore, existing methods also raise the issue of stakeholder involvement; for instance, of justified participation of a relevant group of stakeholders in the design process. It is not our purpose at this point to unfold a detailed elaboration of these shortcomings. Rather, these shortcomings motivate the construction of our method for technology ethics, in the sense that we aim to address them throughout the following sections.

With "method" we mean a detailed, specific problem-solving procedure. However, to this we will add that any method in light of our methodology (i.e., normative theory, developed in the previous chapters) would need to be critical. On the one hand, we should follow Gadamer (2013) in being suspicious of any rigorous procedure that aims at solving an allegedly well-defined problem. As we discussed in Chap. 5, we cannot speak of any *definite* end of a technical practice that cultivates the virtues, which are essentially open-ended with regard to its for-the-sake-of-which, and even less of a formal representation of this end that would allow it to be posited as a definite solution to a definite problem. However, on the other hand we follow Ricoeur in not wanting to face the false choice between "truth *or* method" (Kaplan 2003, p. 34), which would place all considerations of method outside of the scope of our endeavour. Instead, we intend to engage with method in a productive way, making use of its capacities to structure R&I activities while at the same time keeping a critical distance—wary of its power to turn into *dogma*. In order to do so, the procedural steps that we propose will remain open-ended, meaning that they cannot be brought to a definite end but require iterative revisiting, and they will contain explicit elements of critique.

In this chapter, we construct a method for technology ethics starting at the most general, descriptive level, which explicates the central phases involved in the application of the method, and then move to the prescriptive level, which provides more specific guidance regarding how technical practices should be cultivated. First, we outline the method, and present a case of a general technology (personalisation) as well as a specific application (automated borders) that will serve as an illustration that enables us to connect the procedural steps we present with concrete R&I practices. Second, we work out the three phases of our tripartite method: from description to interpretation, interpretation to evaluation and from evaluation to prescription (Table 6.1).

An Iterative, Tripartite Method

To start, the outline of our method needs to be made explicit. At a very general level, our method follows Aristotle's explication of method in the *Nicomachean Ethics,* which Irwin (Aristotle 1999, p. 256) claims to be most clearly stated in Book 7:

> We must set out the appearances, and first of all go through the puzzles. In this way we must prove the common beliefs about these ways of being affected—ideally, all the common beliefs, but if not all, most of them, and the most important. For if the objections are solved, and the common beliefs are left, it will be an adequate proof. (NE, 7.1, 1145b4-8, trans. Irwin)

Table 6.1 Overview of the three phases and nine stages, which together form our novel methods for practising ethics in R&I

Stage:	Procedural steps:
Phase 1: Description < > Interpretation	
I. Map technical practices	Iteratively map technical practices in which a technology is involved
	Consider the different technical practices according to the points of view of making, using, and governing
I.I.**II.** Gather narratives	Select stakeholders, using the categories of users, makers, governors
	Gather second-order narratives
	Gather first-order narratives
I.I.I.**III.** Undertake hermeneutic analysis	Interpret the technology's textuality
	Interpret literacy by considering those using and making the technology
	Investigate the temporality of the process of technological configuration
	Investigate distancing of the process of technological configuration
Phase 1: Interpretation < > Evaluation	
I.**IV.** Uncover standards of excellence	Uncover constitutive rules of a technical practice
	Keep these rules open to critique and revisions
I.**V.** Uncover life plans	Uncover life plans that explain a technical practice
	Explicate life plans according to the movements of ascending complexification and descending specification
I.**VI.** Evaluate narrative unity of a life	Uncover elements that constitute the narrative unity of life
	Establish whether technical practices link to a narrative unity of life or whether one deals with the mere application of a technique
Phase 1: Evaluation < > Prescription	
VII. The good life	Prescribe forms of mentorship that promote self-esteem
	Prescribe forms of an ethical oath, allowing practitioners to gain self-respect through giving themselves a norm
	Prescribe forms of structured dialogue to mediate conflicts and arrive at considered convictions

(*continued*)

Table 6.1 (continued)

Stage:	Procedural steps:
VIII. With and for others	Prescribe forms of community building that promote reciprocity Prescribe forms of codes of conduct that ensure respect for persons Prescribe forms of expert guidance to mediate conflicts in dealing with vulnerable persons and arrive at critical solicitude
I.I.IX. In just institutions	Prescribe forms of civic education that promote equality Prescribe forms of legal regulations that ensure respect for the rule of justice Prescribe forms of democratic decision-making that mediate conflicts between spheres of justice and arrive at a sense of justice

Aristotle argues here that applying the *method* of ethics means firstly to understand the common beliefs about a particular issue, secondly to subject these common beliefs to a philosophical critique, and thirdly to return to these common beliefs by targeting the world of action—wherein the *telos* of any *ethics* lies. Similarly, in line with our discussion of the ethical aim, we argue that any reflection on new ethical concerns that we face due to technological change *starts* from convictions (i.e. common beliefs), which raise certain puzzles. These puzzles pertain to technical practices, for it is the standards of excellence, life plans, and the narrative unity of life linked to practices that are questioned in our ethical framework. In questioning these, we can respond to the puzzles by either confirming or refining the pre-given convictions, which then are turned into considered convictions.

We thereby "map" our method on the three-fold movement developed by Ricoeur: of (1) mimesis 1-practices, (2) mimesis 2-life plan, and (3) mimesis 3-narrative unity of a life, and accordingly, of the move from description, through narration, to prescription. By assuming this tripartite method, we adopt a structure similar to the one proposed in value sensitive design (VSD). However, VSD proposes an order different from ours insofar as it starts with the conceptualisation of values, proceeds with empirical investigations of these values in context, and integrates them in technical investigations that aim at the embedding of values in design (Friedman and Kahn 2002). In contrast to VSD, we do not start with conceptual investigations but with the "empirical investigations", namely with the search for what Aristotle designates as "common beliefs" and what Ricoeur calls "convictions". Only when we have gained an understanding of the narrative structures by which practitioners make sense of a technology and its respective technical practices can we proceed to the conceptual work of criticising these structures. In our method, we additionally distinguish three structural levels: (1) "phase", which refers to the gradual carrying out of a particular task, (2) "stage", which refers to a distinct standpoint in the process of carrying out such a task and (3) "step", which refers to a concrete procedural element of the method.

In its general outline, our method incorporates three over-arching phases, each phase denoting a transition:

1. *Description < > interpretation*: gaining a thorough understanding of a network of technical practices in which a technology is involved.
2. *Interpretation < > evaluation*: interpreting these technical practices according to the puzzles that they raise regarding standards of excellence, life plans and narrative unity of life.
3. *Evaluation < > prescription*: evaluating how these technical practices relate to the good life, with and for others, in just institutions and prescribe ways to improve them.

Illustrating the Method: Automated Borders

Throughout this chapter, we will continually link the development of our method with concrete technical practices in order to illustrate how its procedural aspects relate to the world of action. To this end, we will introduce a specific type of technology as a practical case: personalisation technology. Personalisation technology is involved in an innumerable variety of different technical practices, from detecting fraudulent insurance holders to tailoring learning experiences for children in primary schools. To also consider a more particular technology, as in a type of technical object, we will additionally narrow our focus on the case of smart, or automated border control systems that incorporate the technical process of personalisation.

Personalisation can be applied to a wide range of different technologies, but it generally revolves around notions of "adapting", "fitting" or "tailoring" digital content to the human being(s) interacting with it. A central term in personalisation research is said to be the "adaptability" of a system (Lycett 2013). Adaptive systems include three basic models in their design: a "user model" that is a structured model for the collection and categorisation of personal data belonging to a user, "an application model", which is a description of relevant features of the application, and an "interaction model" that is meant to structure the organisation of interactions between a user and a system (Lycett 2013). Asif and Krogstie argue that multiple personalisation approaches exist that can be based on "machine-learning algorithms, agent technology and ubiquitous and context-aware computing" (Asif and Krogstie 2012, p. 346). They identify a "basic level" of personalisation, at which a user selects a certain configuration of a computing device or interface, which subsequently remains the same. This basic level corresponds with the conventional, instrumental view of a personalisation technology: of the human user determining the technology's settings and behaviour. Then there is a "second level", at which the configuration of a system is based on a user's "profile", and a "third level" at which both the profile of the user and his or her "context" (comprised of metadata such as location, time of the activity, type of activity) are used as the basis for the configuration of the system. Bouzeghoub and Kostadinov (2007) make a distinction between profiles and queries: a profile being a "user model" "defined by a set of attributes" and a query being an "on-demand user need". Roosendaal (2014) offers an additional distinction, namely one between "digital personae", which are representations of known individuals in the real world and "digital profiles" which are sets of characteristics about persons that can be used as inputs for algorithmic decision making.

The process of personalisation invariably integrates underlying implicit or explicit assumptions about *what* a person is and—more importantly—about what we can understand as the *right* information for the *right* person. It is an automated process of categorisation, and therefore of inclusion and exclusion of both digital content and of digital personae or profiles from the information people access. As a result, personalisation processes can for instance influence what type of information we retrieve from our search engine, what kinds of products are recommended to us when browsing online, and what kind of feedback we receive about our daily activities. Also, personalisation processes are used by intelligence agencies to identify suspicious individuals and by insurance companies to establish people's personal credit ratings. Accordingly, these increasingly ubiquitous computational processes that directly influence many aspects of people's everyday lives can have significant ethical implications. For instance, de Vries (2010) argues that ethical concerns arise from cases of implicit discrimination based on profiling. Moreover, Bohn et al. (2005) argue that the practice of matching a digital persona with profiles that present a public security concern has ethical implications because it can lead to mass-surveillance practices. In a different vein, Schubert (2015) argues that the use of personalised technologies that nudge their users in certain directions can lead to a reduction of personal autonomy and agency.

A concrete and rich example of personalisation based on a digital persona and profiles is an automated passport check, or smart border control, at an airport (Juels et al. 2005). When someone's passport is scanned, the retrieved data is compared with a data entry containing the document number belonging to the respective person, accompanied by her picture, biometric information and information about country of origin, age, and so on. Based on an algorithmic assessment of this personal data, the person can either pass through or will be held and interrogated by the border police. In this process, the digital persona can be compared with and transformed into a digital profile, for instance by linking it with a certain profile containing passport features that are deemed "suspicious", or by using it to add to the digital profile of people originating from the same country. What the process of personalisation in smart border control systems has in common with personalisation processes in general is that it uses interactions with humans to gain knowledge about them and create a representation that changes the behaviour of a system in order to fit, adapt to or be tailored to this representation. The purpose personalisation research is to make sure that the behavioural changes of the system approximate the expectations, wishes and needs of the human interacting with it. Terms like tailoring suggest that the user is seen as a given, as a static point to which the personalisation processes need to adjust. Just like a tailor adjusts the sizes and shapes of a piece of clothing to the human body that is supposed to remain the same (or rather, that is defined by fixed measurements), personalisation processes are supposed to be tailored to users who are presumed to remain the same. Accordingly, user needs or preferences are supposed to be fixed. For instance, it might be assumed that the user of a weather application has a fixed need of knowing what the weather will be the next day in her city, and on request a personalisation process will link the data of her location and weather forecasts to provide for the desired information.

In what follows, we will usually refer to personalisation technology in general, and smart border controls in particular, to illustrate the use of our method. That is, we will relate the technical practices that we need to consider in our method to personalisation technologies, and use them to explicate relevant stakeholders, basic actions, narrative structures, and so forth. We will present a number of fictional accounts of narratives related to technical practices in which smart border control systems are involved. Occasionally, however, we will also illustrate some points with reference to familiar technologies different from smart border control systems (e.g., smart phones). The fictional accounts we provide are to be taken solely as illustrations, explicating how the method could apply to specific technical practices, and are not to be taken as demonstrations that could be part of an actual case study.

Phase I: Description < > Interpretation

The first phase of the method concerns the move from a description of technical practices to their interpretation. Suppose one is a philosopher who is charged with applying our novel method to a particular R&I project that aims at developing a personalisation technology for smart border control systems. In accordance with our methodology, the first step would be to uncover the different technical practices in which such a technology is involved, which can be categorised according to the making, using, and governing of a technology. The second step to take would be to uncover the narrative structures that characterise one or more technical practices. This can both be done by (1) gathering narratives *about* a technology, and by (2) investigating the relevant process of technological configuration (i.e. focusing on the design of the technology). These narrative structures can consequently be used as the basis for the hermeneutic analysis that is informed by the narrative technologies approach, and which aims at explicating the textuality and literacy concerning the process of technological emplotment and the temporality and distancing in the process of technological configuration.

Mapping Technical Practices

The first stage is concerned with the mapping of technical practices to uncover a network of technical practices in which a particular technology is involved. The uncovering of such a network is in a way similar to uncovering the cluster of actants in actor network theory (Latour 2013), but instead of focusing on actants and the semiotic relations between them it focuses on a cluster of technical practices and their hierarchic organisation according to constitutive rules, life plans and the narrative unity of a life. We need to emphasise what sets our approach, grounded in virtue ethics, apart from most other methods for practising ethics in R&I: namely, the focus on *technical practice*. That is, our methodology shows that what is at stake in any ethical appraisal of a technology is neither merely the thing in question—the smartphone, test tube, self-driving car, and so forth—nor its potential consequences—existential risks, climate change, the death of a patient, and so forth—but foremost the technical practices in which the technology is involved. This is not to say that the things or their consequences do not matter, but simply that these only matter *with regard to* a technical practice and not considered in isolation. Thus, if we for instance want to evaluate the ethical impacts of a smartphone, we should not focus on the thing, but on the technical practices in which it is involved, such as "having an online conversation".

Now, we need to address two important objections to such starting point. First, one could object that not every interaction with a technology constitutes a technical practice as we earlier conceptualised it. For instance, we could hardly speak of the ethics of "swiping" or "checking the weather" on a smartphone, because these activities do not relate to standards of excellence and life plans. Yet, these activities should not be considered ethically "neutral", for they could bring about benefits or harms, so we should include them in our considerations. As a response, we reply that indeed what is at stake when considering these activities would need to be extended to a technical practice in order to be ethically significant. As such, we cannot speak of "swiping" on a smartphone as being either good or bad or bringing about good or bad character, even when we could establish that excessive swiping could for instance lead to health problems. When the latter is the case, we ought not to focus on the swiping as such, but on the technical practice in which it is involved, such as "playing video games". Such a technical practice links to standards of excellence and to life plans (of leisure time). Addressing the basic activity of swiping means therefore primarily addressing technical practices such as playing video games on a smartphone, which can explain *why* a smartphone user engages in excessive swiping.

Second, one could object that a technology can be involved in innumerable technical practices, which makes it impracticable or even impossible to focus on technical practices instead of on the thing in question, or on its consequences. This brings us to the problem of uncertainty: the further a technology is removed from any concrete application, the less we can reliably speak of a technical practice in which it is involved. For instance, R&I of quantum computing is still in a very experimental phase and we cannot yet say anything reliable about the way in which it will play a role in technical practices of those who will use it. We can therefore speak of a strong indeterminacy of technical practices when considering R&I of quantum computing—and for that matter of most emerging technologies. As a response, we argue that indeed we face indeterminacy with regard to technical practices of the *use* of quantum computing. In the words of Ihde (1979), we could say that the *use context* of quantum computers is thus far indeterminate. However, we do not face the same indeterminacy regarding the technical practices concerned with *making* quantum computers (designing hardware components, experimenting with software implementations, etc.), because these clearly link with standards of excellence and life plans. Similarly, we face significantly less indeterminacy with regard to technical practices concerning the *governing* of quantum computing R&I (e.g., curating research results, securing research facilities, etc.). We therefore argue that though we need to embrace the indeterminacy of technical practices of use when dealing with emerging technologies, we should focus on technical practices related to the *making* and *governing* of these technologies. In other words, not yet having sufficient knowledge concerning the use of emerging technologies does not preclude efforts to shape technical practices that cultivate the virtues of making and governing these technologies.

How could the mapping of technical practices be done? First, we need to stress that this mapping is an *iterative* task: that new relevant technical practices can be uncovered on the way, as new narrative structures are taken into account. For instance, a researcher might at some point provide an account of a new type of use of a technology that was not thought of before, and can thereby add this technical practice to the network. Second, each technology should be considered regarding technical practices of *making, using,* and *governing,* and not merely with regard to use. Even though these types of technical practices all have the same structure (for instance, both the *making* of software for a smartphone and the *use* of a smartphone to have a conversation have their related standards of excellence and life plans), they relate differently to the technical object in question. Making and using refers to technical practices that involve the first-order technological configuration of a particular technology. *Making* refers to technical practices concerned with the being-completed of the technical object, for instance designing the camera for a smart border control system. *Using* refers to technical practices concerned with the in-order-to of a technical object, for instance putting one's passport in the smart border control system in-order-to cross the border control. *Governing,* then, refers to the broad category of technical practices involved in the second-order technological configuration of a particular technology (i.e. concerned with narratives *about* the technology), for instance technical practices that regulate the collection and curating of personal data at the smart border control. Governing thereby does not merely refer to technical practices of domination (e.g., making laws to govern the use of a technology), but also to practices of power-in-common (e.g., online protests against laws that aim to govern the use of a technology). Even though the types of *making, using* and *governing* often overlap (e.g., in *making* a camera, an R&I practitioner *uses* a 3D printer), these types of technical practices need to be distinguished because they enable one to uncover technical practices in which a technology is involved from three distinct points of view.

We can formulate the first set of procedural steps of our method, aimed at gathering together a network of technical practices:

- Iteratively map technical practices in which a technology is involved.
- Consider possible technical practices according to the points of view of *making, using,* and *governing* this technology.

Gathering Narratives

The second stage is concerned with gathering narratives of particular technical practices. Once different technical practices have been identified, they need to be made explicit. That is, in order to *understand* the technical practice, one needs to know the narrative structures through which it can be explained. These narrative structures are grounded in what Aristotle calls common beliefs. This implies that a selection of stakeholders has to be made, which might include any practitioner who engages in the technical practices making, using, or governing a technology. For instance, in the case of smart border controls, users are people passing through these passport controls in airports and other places where border controls are implemented. Makers can be engineers contributing to the design of the system, the software for the system, or building the infrastructure needed for the system. The people engaging in technical practices of governing can be government officials in ministries of foreign affairs, or members of parliament who propose legislation specifically targeted at regulating smart border controls. When considering the category of governing, peripheral stakeholders should also be taken into account, meaning those stakeholders that have little to say about a particular technology but that do have a "stake" in the way it is made and used. For instance, people who don't use social media (i.e. "non-users") have a stake in the use of social media because of the so-called network-effect, meaning that by not using social media they can be left out of significant social communication (Hargittai 2007). As with the mapping of technical practices, the selection of stakeholders is an iterative process, meaning that new stakeholders can be encountered along the way.

In order to uncover the *prefigured time* of technological emplotment with regard to particular practices, which implies gathering second-order narratives *about* a technology, one can look at different established sources, which can come from the media, academic opinion pieces, literature, or other significant spaces where stories are expressed such as the "blogosphere" on the Internet (Adamic and Glance 2005). Arguably, philosophy cannot provide any strict in- or exclusion criteria for the narratives to be considered, because the much-contested criterion of relevance would need to be the differentiating factor (e.g., relevance according to popularity, expert opinion, and so forth). We will therefore not determine any particular way of establishing relevance and argue that practitioners should venture outside of philosophy to find proper tools to do so (e.g., trend analysis or bibliometric analysis). Again, however, it makes sense to look for technical practices of making, using and governing, amongst other reasons because narratives concerning technical practices of using are often unavailable when considering an emerging technology. In such a case, it is more valuable to look for instance at technical practices of making, such as the ones explored in an article about the working environment in Silicon Valley tech companies (Weigel 2017). In the case of smart border controls, which are already being implemented, we can for instance consider the comment piece in Nature about borders as a source of second-order narratives (Clavell 2017). This piece offers accounts of technical practices of using (e.g., "traveling") and of governing (e.g., "monitoring"), and also provides certain standards of excellence, for instance for monitoring that it should involve "a rapid and straightforward verification" and that it "should not discriminate" (Clavell 2017, p. 35).

In order to subsequently uncover the *configured time* of technological emplotment (concerning technology as a co-author of narrative structures), one needs to gather narratives that recount the engagement in a technical practice, for instance narratives of someone recounting queuing for passport control, walking into the booth of a smart border control system, sliding the passport into the system, waiting for one's face to be recognised, and so forth. In other words, one should gather first-order narratives, coming directly from those engaging in a technical practice. In the case of smart border controls, one would want to gather narratives concerned with the process of technological emplotment from makers (e.g., the people assembling the technology or designing its infrastructural support), users (e.g., travellers moving through international airports) and governors (e.g., border police officers operating the systems). Because the process of technological emplotment relates to the *design* of a technology (i.e., design features that mediate the temporality and distancing in the process of emplotment), first-order narratives should involve aspects of design, which implies that the person gathering them should ideally have a certain level of knowledge of a particular technology design. When considering emerging technologies, it is especially crucial to involve makers (i.e., R&I practitioners), because they have the earliest and most well developed access to the first-order narratives in the process of technological emplotment (cf., Brey 2000).

At this point, we can formulate the second set of procedural steps, concerned with gathering the narratives that will allow for the hermeneutic analysis:

- Select stakeholders, iteratively, and according to the three overall categories of *users, makers,* and *governors*.
- Gather second-order narratives from sources that present stories *about* a technology.
- Gather first-order narratives from stakeholders engaging with the process of technological emplotment.

Hermeneutic Analysis

Once the relevant narratives concerning technical practices have been gathered, they should in the third stage be analysed according to the narrative technologies approach. This practice of analysing a technology incorporates the four hermeneutic concepts of *textuality, literacy, temporality,* and *distancing*, and uses them (as textual metaphors) to interpret a technology-in-practice. The interpretation of a technology according to each of these concepts depends on the human interpreter, but can be structured according to a number of steps. First, concerning *textuality*, an interpreter should look at the differences between prefigured and configured narrative structures. For instance, if a narrative about a bridge (e.g., recounting that the bridge enables the trade from one side to the other side of the river) is similar to the recounting of the actual use of the bridge (e.g., "I cross the bridge every day to deliver my goods to the company at the other side"), the interpreter could conclude that the process of emplotment involves a low degree of textuality. Conversely, if the narrative about a technology (e.g., recounting that driving a car implies following certain traffic rules in particular situations) and the narrative about the use of a technology (e.g., "I needed to change gears and drive in a lower speed because of bad visibility at that particular corner"), are distinct, the interpreter could conclude that the process of emplotment involves a higher degree of textuality. Here, the situation and the technology play a much more active role: there is emplotment by the technology in a particular situation. This difference can also be established through inquiries into cultural references that imply a distinction between users and non-users, such as those implying the distinction between digital natives and non-natives, which points at a high level of textuality. The degree of textuality shows the interpreter *where to look* and focuses on the question regarding agency (by humans and/or technology): whether to merely investigate narratives *about* a technology or also narratives concerning the technological emplotment *by* a technology.

Second, concerning *literacy*, an interpreter should consider the types of stakeholders, relating to the making, using and governing of a technology. For instance, in the case of nuclear weapons, only few makers (e.g., nuclear scientists) and users (e.g., military personnel with high-level clearance) can be considered, but many people can be considered engaging in practices related to the governance of these technologies (e.g., members of the International Atomic Energy Agency, people in protest groups, members of parliament, etc.). From this, one could conclude that the degree of literacy concerning nuclear weapons is low. Conversely, when considering smart border controls, many users and a considerable number of makers can be considered, which suggests a high degree of literacy. The degree of literacy shows the interpreter *to whom to look* and focuses on the question concerning knowledge (of social actors): whether one should engage with experts or rather more with non-expert stakeholders. In order to further investigate the process of technological *configuration* (i.e., the mediating stage between prefigured and refigured time), an interpreter should focus on its dimensions of temporality and distancing. These dimensions relate to technical practices in each case. For instance, in Chap. 4 we mentioned aspects of the design of a car as contributing to the temporality of the technical practice of driving. Similarly, we argued that design features of electronic derivative trading systems contribute to the abstraction from the world of action in the technical practice of derivative trading.

Concerning the *temporality* of the plot, an interpreter should focus on the way in which design features organise human action either according to a chronological sequence of actions or according to a non-chronological ordering of actions. Here the question is *when* happens what, in what order. It is a question concerning time. As we saw in Chap. 4, she can thereby focus on the succession of events or on the presence of connectors (i.e. dating mechanisms, version control mechanisms, tracing mechanisms). In the case of smart border control systems, for instance, an interpreter could focus on the following account provided by a traveller: "I first had to walk through the little glass door, stand on the two footprints on the floor, and place my passport in the machine. I then had to look at a certain point to have my picture taken". Similarly, a border control officer could recount: "each passing traveller gets a unique ID connected to the time-stamp of border crossing". These accounts indicate that regarding the technical practice of "border crossing", the smart border control system predominantly configures the chronological dimension of time, enforcing a strict sequence of actions.

Concerning the *distancing* of the plot, an interpreter should aim at uncovering crucial entities and events that are either close to the world of action or abstracting from it. This is another "where" question, but this time it asks for the degree of abstraction from the world of action. As we argued in Chap. 4, abstraction leads to a process of *forgetting*, for instance forgetting that through algorithmic trading of a certain type of electronic bond, house owners might be affected. To revert this process, one has to start from the abstracted level at which quasi-entities and quasi-events operate. In the case of smart border controls, quasi-entities are for instance digital profiles that result from the process of abstraction of concrete individual travellers into aggregates of demographic profiles of these travellers. By investigating the origin of these quasi-entities (i.e., by referring-back to the world of action), the interpreter will be able to uncover how these entities affect actual people engaging in technical practices. For instance, when a traveller with a certain physical appearance is stopped at the border, this could be explained by referring-back from the technological configuration involving an abstracted profile (e.g., representing the aggregate of properties of a "suspicious individual") to the world of action in which an actual person is affected.

The third set of procedural steps applies the narrative technologies approach, and offers the interpretative framework supporting the evaluation of technical practices:

- Determine where to look by interpreting the textuality of a technology, focusing on the difference between narratives about and by technological emplotment.
- Determine to whom to look by interpreting the literacy of a technology, focusing on the difference between stakeholders relating to the process of technological emplotment.
- Investigate the temporality of the process of technological configuration, focusing on the organisation of events and the presence of connectors.
- Investigate distancing in the process of technological configuration, referring-back from quasi-entities and quasi-events to the world of action.

Phase II: Interpretation < > Evaluation

The second phase of the method concerns the transition from interpretation of technical practices in accordance with the narrative technologies approach to their evaluation in accordance with the narrative unity of life. In line with the theory of technical practice that cultivates the virtues as laid down in Chap. 5, this first involves interpreting the move from basic actions to technical practices, according to standards of excellence. Secondly, it involves the interpretation of life plans, as those movements that make technical practices intelligible. Thirdly, it involves a consideration of the narrative unity of life to which technical practices and life plans are related.

Standards of Excellence

The fourth stage concerns the uncovering of standards of excellence, which again is an iterative exercise because new standards of excellence can be uncovered on the way. For this exercise, the hermeneutic analysis should be used. First, the degree of textuality indicates whether to take design into account in considering the standards of excellence or whether to take narratives into account *about* a technology. Considering smart border control systems, we will have to inquire into the process of technological configuration to uncover certain standards of excellence. Second, the degree of literacy indicates which stakeholders can provide valuable insight into the existence and contents of standards of excellence. For instance, it indicates that travellers going through airports are important stakeholders to consider with regard to smart border control systems. Third, temporality indicates the degree to which standards of excellence imply a strict succession of events or a course of action that can be freely chosen. In the case of smart border controls, most standards of excellence of the use of such systems relate to the strict succession of events. Fourth, distancing indicates the degree to which standards of excellence might be forgotten. Considering smart border control systems, this might imply that by referring back from the abstract profiles that are used by border police officials to concrete individuals, decisions based on these profiles could imply a forgetting of the standard of excellence of treating each passing traveller according to her personal circumstances.

We need to recall that standards of excellence are initially considered to be constitutive rules, which are not yet strictly speaking "ethical" and therefore merely provide a basis for the way technical practices can be evaluated. For instance, a constitutive rule for the technical practice of "crossing the border" might be "walking into the booth of the smart border control at a regular pace counts as a valid way to initiate the process of border crossing". To uncover constitutive rules related to practices, relevant stakeholders should be asked about the way in which they perform certain technical practices. Accounts of these standards of excellence can already be found in the narratives that were gathered in the first phase, but could additionally be gained from focus groups (Bloor et al. 2001) or narrative interviewing (Jovchelovitch and Bauer 2000). Channelled questioning of the way of the technical practice of border crossing could for instance lead to an account like the following: "in crossing the border, I try to keep calm and move through the border control quickly, not to keep up other people in line". From this, one could derive "keeping calm" as part of a constitutive rule that mediates the technical practice of border crossing.

It is important to note that one should not yet take constitutive rules as properly "ethical" standards of excellence. In many cases, constitutive rules can actually obstruct the cultivation of virtues, for instance a constitutive rule such as: "looking away counts as a valid response when witnessing an assault in a public space". However, constitutive rules add to the understanding of "common beliefs": to the everyday understanding of a technical practice. For this reason, one should keep the constitutive rules *as* standards of excellence *open to critique and revision* in a social and practical context, i.e., the relevant social practices and political communities, where the ethical aim can be invoked to change the constitutive rules. For instance, later considerations of the ethical aim could lead to a transformation of the above-mentioned standard of excellence into: "gathering support and attending to a victim counts as a good response when witnessing an assault in a public space".

The procedural steps that can be derived from this initial consideration of standards of excellence are the following:

- Uncover constitutive rules of technical practices in question, by asking relevant stakeholders about the way they engage in these practices.
- Keep constitutive rules as standards of excellence open to critique and revision.

The latter is not just the work of philosophers; this work needs to be done in the community of practitioners and stakeholders. Moreover we could link here to participative democracy theories and to theories and practices of responsible research and innovation, in so far as such frameworks demand that stakeholders are involved in such a critique and revision.

Life Plans

The fifth stage concerns the uncovering of life plans, or the movements of emplotment through which technical practices are made intelligible. As discussed in Chap. 3, life plans are characterised by the movements of ascending complexification and descending specification. These provide explanations of *why* agents engage in certain technical practices. It is important to note that new standards of excellence can be uncovered in the consideration of life plans. This can be illustrated by discussing the notion of integrity that comes along with being a border police officer, which might be expressed in a narrative such as: "I'm working at the border control because I chose to be a police officer and serve the public interest". Related to this life plan, standards of excellence for the configuration of smart border controls could emerge such as: "the system should not discriminate between people of different ethnic backgrounds".

To uncover life plans, the movements of ascending complexification and descending specification need to be made explicit. To do so, an investigator should inquire how narrative structures that constitute life plans include accounts of how basic actions explain more or less remote ideals and vice-versa, how more or less remote ideals explain basic actions. For instance, a person going through the border control could recount: "I had to keep standing in place and look at the camera while it tried to capture my face without success. This experience violated my sense of justice". Such a statement would indicate a link between basic actions (i.e. standing, looking) with the remote ideal of justice. Conversely, an engineer might explain: "I recently had a discussion with my colleague about privacy, after which we decided to change some settings of the smart border control, giving border control officers less insight into irrelevant details of travellers once they touch the screen to prompt a traveller's profile". Such an account links the remote ideal of privacy with basic actions like touching the screen.

We derive a fifth set of procedural steps from the discussion of life plans, focusing on the movement of emplotment:

- Uncover life plans of technical practices in question, by asking relevant stakeholders to explain why they engage in these practices.
- Explicate life plans through the movements of ascending complexification and descending specification, focusing on basic actions that are explained through more or less remote ideals, and vice-versa.

Narrative Unity of Life

Engaging in a preliminary evaluation of the narrative unity of life constitutes the sixth stage. This evaluation is primarily aimed at understanding whether technical practices nest their in-order-to with regard to their for-the-sake-of-which. In other words, it establishes whether a technical practice can be linked with a notion of living together in a political community, which is necessary for it to be qualified as virtuous. Looking back at our discussion of Heidegger in Chap. 3, we state that this step allows us to differentiate between the mere application of a technique and a technical practice. More strongly and in the light of moral questions, we could state that this difference denotes the distinction introduced by Arendt in her later work, between the "banality of evil" (Arendt 1964) and "acting viciously". That is, even before the possibility of acting virtuously or viciously is opened up, there is an "inability to think" (Ezra 2007, p. 141) in the mere application of a technique. This notion coincides with what Heidegger calls the "supreme danger" of technology (Heidegger 1977, p. 26), and for both Arendt and Heidegger this implies that the in-order-to of a technical activity is indefinitely extended. In other words, whenever a practitioner is unable to link the in-order-to of her technical practice with its for-the-sake-of-which, she is unable to act virtuously or viciously but is able to engage in banality, which is dangerous considering the growing power of technology. Having a narrative unity of life that supplies one with the for-the-sake-of of a technical practice is therefore necessary for such a practice to be called "virtuous".

To uncover the "unstable mixture of fabulation and actual experience" (Ricoeur 1992, p. 162) that constitutes the narrative unity of life to which a practitioner can appeal when recounting technical practices and life plans, one should focus on elements of the narrative structures that recount how one wants to live his or her life. As such, this narrative unity is the basis for a naïve understanding of the good life. As can be derived from Ricoeur's characterisation of the narrative unity of life, it is likely to appear as a fragmentary structure that draws from different subordinate elements, which might be standards of excellence or life plans. For instance, an employee of a civil society organisation that criticises the use of smart border controls might state the following: "I was visiting my sick grandmother when the system kept me waiting in the airport for no apparent reason, which caused me to arrive too late. This experience led me to oppose the use of smart borders, because they violate my idea of how people should treat one another in this country. I believe everyone should be able to appeal to a person's judgement when having to pass an arbitrary border." Such an account contains the recounting of a life plan and constitutes what might be called a fragment of a narrative unity of life, for it supports one's practice by appealing to a notion of living together in a political community. Even though this initial notion of the narrative unity of a life does not yet offer any grounds for saying *how* a technical practice contributes to the good life, it does indicate *whether* a technical practice relates to a notion of the good life, or whether we are dealing with a mere application of a technique.

The discussion of the narrative unity of life provides the method with a sixth set of procedural steps, which focuses on the mixture of fabulation and actual experience that makes up the narrative unity:

- Uncover elements that constitute the narrative unity of life, by inquiring into accounts of living together in a political community.
- Establish whether technical practices link to a narrative unity of life, or whether one deals with the mere application of a technique.

Phase III: Evaluation < > Prescription

The third phase of our method is concerned with the move from an evaluation of the narrative unity of life and a consideration of the political community as a context to *prescription* in accordance with the ethical aim, which establishes a *stable* for-the-sake-of-which of technical practices. This exercise leads to a basis for prescription because it allows for the establishment of grounds (i.e., reasons) for imputation of a person with regard to her actions, of responsibility with regard to others, and of recognition with regard to unknown others. In other words, after this final phase we can formulate prescriptions for an agent to engage in technical practice that cultivates the virtues concerning the extent to which this practice supports the good life, with and for others, in just institutions. In this phase, we explicitly reconnect with the heuristic of virtues that we presented in Chap. 5. We will do so by connecting the self-regarding virtues (e.g., courage) in that heuristic to an understanding of the good life, other-regarding virtues (e.g., friendship) to an understanding of "with and for others" and virtues concerning unknown others (e.g., justice) to an understanding of "in just institutions". Practical forms of prescription based on considerations of the ethical aim might take many different forms and do not only refer merely to deontic categories of obligation, permission, or prohibition. In addition, prescription might take the form of educational guidelines, design specifications, constitutional norms, procedures for just deliberation, and so forth.

In accordance with the explication of the ethical aim in Chap. 5, each stage of phase 3 will contain a threefold structure: starting with the naïve conception of the good life and norms given within a particular practice and community, moving through the sieve of the norm, and reflecting on the norm on the basis of a refined notion of the ethical aim. The first segment of each stage will be concerned with the "naïve" conception of the good life that serves as the narrative unity of life in relation to technical practices. This conception is "naïve" insofar a practitioner has not yet reflected on her intuitive understanding of the good life, and thus has not yet subjected her understanding of the good to its limitation by norms. This conception of the good life makes sure that we are not dealing with a mere application of a technique but with a genuine technical practice that incorporates standards of excellence and life plans in relation to a narrative unity of life. The second segment of each stage will be concerned with the formulation of the basis for a normative structure that might govern a particular technical practice. This normative structure ensures that the possibility for evil is minimised. The third segment of each stage will be concerned with the formulation of forms of arbitration and will therefore underlie grounds for shaping *participation* in practising ethics in R&I. As such, it presents a basis for (1) *how* participation should take place, (2) *who* should participate, and (3) *how* participation can be legitimised. As such, participation is the cornerstone of the method to arrive at considered conviction, critical solicitude and a sense of justice.

We need to emphasise two aspects of the way stakeholders are considered in our method. First, "participatory design" in the context of our method does not merely mean that people design a technology together, but more importantly that people have a recognised voice in how a technology should be designed. This implies that participatory design cannot be accomplished without democratic governance that offers participants a basis from which to question design decisions, to search proper guidance in making design decisions, and to appeal to just, democratic decision-making procedures. Second, even though we respond to the issue of "value conflicts", the conflicts that the method deals with—between autonomy, respect, institutions, and the refined ethical aim—are explicitly *not* value conflicts, but conflicts between conceptions of the good and obligations, or between norms and the good life. As such, it is not values—which Ricoeur designated as being "compromise terms"—that conflict, but actualisations of conceptions of the good life in norms, laws and institutions and the tension between these and a reflection back on a better informed notion of the good life that is the basis for arbitration to solve conflicts.

The Good Life

The seventh stage offers ways for reflecting on technical practices that link these practices to a notion of *the good life*. At this point, we reconnect with the heuristic of virtues offered in Chap. 5, because they offer points of entry for thinking about self-esteem. We will consider prescription of different aspects of technical practice, according to Ricoeur's notion of the good life in terms of self-esteem, self-respect, and considered conviction.

Self-esteem. This refers to the extent to which a practitioner can evaluate her actions in relation to a naïve conception of the good life. This is where the virtues of character come into play, which pertain to individual conduct. In line with Vallor's (2016) heuristic of virtues, we argue that a practitioner needs to possess the self-regarded virtues of *humility, courage, perspective, self-control* and *magnanimity* in order to gain self-esteem. Our understanding of a technical practice that was developed in phases 1 and 2 assists in understanding how a technology either obstructs or supports the cultivation of these virtues. For instance, regarding magnanimity, we could argue that because of the strict enforcement of the sequence of events, a traveller is obstructed in showing leadership when this might be needed while engaging in the technical practice of border crossing. For instance, she will be less inclined to encourage bystanders to object to the treatment of a fellow traveller, both because she faces a strict succession of events that she cannot reasonably interfere with and because this precludes her to appeal to another person (i.e., a member of the border police), whose leadership she might call into question. As such, smart border controls could obstruct her sense of self-esteem while engaging in the technical practice of crossing the border.

The primary forms of prescription in relation to self-esteem are forms of personal development or *mentorship*. For instance, as we already alluded to in Chap. 5, human resources departments, which might better be called human development departments, play an important role in making sure R&I practitioners can relate the technical practices they engage in to a naïve conception of the good life. For instance, they aim to ensure that people enjoy the work they do for the good reasons (i.e., not merely for monetary gain but also for the conviction that doing the work is the right thing to do). In common words, one could say that mentorship should lead practitioners to have a sense of purpose in what they do, which is the bare minimum for engaging in a technical practice that cultivates the virtues.

Self-respect. This refers to the extent to which a practitioner evaluates her actions in relation to a norm that she gives to herself. This norm derives from the necessity to limit one's actions in accordance with what Ricoeur would call the facticity of evil. Consider for instance the movement of descending specification of a border police officer having a conversation with his colleague about privacy, leading him to change the settings of the smart border control system. By doing so, he gives himself a norm according to which he could evaluate the standards of excellence of acting while being a police officer, which in relation to the naïve conception of the good life relates to protecting the public. In this context, "privacy" can therefore be understood as the norm that limits "security" as a public good.

The primary form of prescription of the norm is that of a *code*. As such, the norm is for example reflected in what used to be Google's first rule in its corporate code of conduct, "don't be evil" (Miller 2005). Even though our method provides a good basis for actually criticising this rule for not being *sufficient* for ensuring technical practices that cultivate the virtues, the formulation of this rule does illustrate the basis for the norm leading to self-respect. However, in opposition to the Google code of conduct, we need to emphasise that the norm has to be *given* to oneself in order to ensure its enacting in technical practices. This is why we referred to the importance of an *ethical oath* (Blok 2013) as the proper means for ensuring a practitioner to give the norm to herself. As such, an ethical oath functions as what Ricoeur designates as a "test" (Ricoeur 1992, p. 224), a means for a practitioner to test whether her naïve conception of the good life invoked in a technical practice does not violate the norm.

Considered conviction. This refers to the extent to which a practitioner is capable of dealing with conflicts between the naïve conception of the good life and the norm she gives herself. Responding to these conflicts requires engaging in finding a reflective equilibrium between an ethics of argumentation and considered convictions that are historically situated. Consider for instance a smart border system that not only lets people halt, wait or pass, but also engages in a close-up investigation in search of dangerous or illegal items on a person's body. An engineer might implement a rule that reflects a norm stating that such an automated investigation should be carried out on the grounds of automatically detected suspicious aspects of a person that in no case link up with a person's ethnicity or gender, which should therefore not be recorded. The norm reflected by this rule would therefore limit the technical practice of "securing the border" by implementing a restriction of non-discrimination. However, in a particular case a woman from a certain ethnic background might be harmed by the implementation of this norm because she assigns the male gender to robotic entities and because of her cultural background sees it as immoral to be intimately investigated by a male. In such a case, a conflict arises between the norm and historically situated convictions.

The most important form of prescription following from the consideration of conviction is one of *structured dialogue*. Since the above-mentioned conflict happens in public life, its resolution depends on the capacity and inclination of different practitioners to engage with one-another. On the one hand, considered convictions need to be taken into account, which for instance means that the design of a smart border control system should allow a user to override its automated sequence of events and request a dialogue with a human. The person responding (i.e., a member of the border police) could structure the dialogue according to principles as laid out in discourse ethics, which would allow her to consider the traveller's conviction while also upholding the interest of the public, ensuring a sufficient level of security. In the case described above, a structured dialogue might for instance lead to the decision to have a female agent perform the security check.

The above-mentioned discussion dealing with the good life provides us with the following, seventh set of procedural steps:

- Prescribe forms of mentorship that promote the relevant virtues of humility, courage, perspective, self-control and magnanimity, which incline practitioners to have self-esteem while engaging in a technical practice.

- Prescribe norms in the form of an ethical oath, which allows practitioners to gain self-respect by giving themselves those norms in a public setting.
- Prescribe principles for structured dialogue, which should mediate conflicts between norms and historically situated convictions, to arrive at considered convictions.

With and for Others

The eighth stage offers ways to reflect on the way in which technical practices link up with a notion of the good life, not merely for the individual worthy of esteem but also with and for others. Again, we start by considering the naïve conception of "with and for others" in the notion of reciprocity, by integrating the other-regarding virtues developed by Vallor. From this initial notion of reciprocity, we move through the sieve of the norm of respect for persons towards the reflective notion of critical solicitude.

Reciprocity. This refers to the extent to which a practitioner interacts with others (e.g., with colleagues, customers, and so forth) in relation to a naïve conception of the good life *with and for others*. The other-regarding virtues of character allow us to evaluate a technical practice in this regard, which in Vallor's approach are *care, friendship* and *honesty*. We encounter these virtues in Ricoeur's work as well, where the virtue of friendship initiates the discussion concerning the other-regarding character of the ethical aim, where honesty comes into play while considering the virtue of "keeping one's word" (Ricoeur 1992, p. 118) as being constitutive of selfhood, and where care is reconceptualised as solicitude. We can relate these other-regarding virtues of character with an analysis of technical practice. For instance, we could argue that when a border policy officer reconfigures the technical practice of "crossing the border" after having had a discussion with his colleague about privacy, the other-regarding virtue of friendship is at stake. That is, the friendship that binds him with his colleague inclines him to engage with the for-the-sake-of-which of the technical practice that they together engage in, which are linked together in one or multiple life plans.

The principal form of prescription to establish the most basic form of reciprocity is what in management literature is usually referred to as "team building" (Liebowitz and de Meuse 1982), but which in broader society can be captured instead with the notion of "community building" (Lichterman 1995). Both these types of prescribed activities are focused on cultivating the other-regarding virtues. They thereby contribute to the bare minimum for people engaging in technical practices together to do so with and for other people.

Respect for persons. This refers to the extent to which practitioners inter-act in a technical practice while treating each other never merely as a means, balancing agency and patiency. This places a limit on technical practices, by making sure through the inauguration of a norm that respect for persons will be upheld. We can understand how this actualisation of respect through the norm comes about by considering our understanding of a technical practice. Considering smart border control systems, we could argue that when faced with the question of automation, engineers ought to take into account the involvement of human border police officers in securing the border. Balancing their agency and patiency would mean that their voices ought to count in making a decision on automation instead of reducing such a decision to mere efficiency considerations, which would reduce practitioners to mere means. This would lead to a clear choice between technological alternatives, where a symbiosis between man and machine would be preferred over full automation, while considering a particular technical practice of securing the border.

Since we again deal with a *test* when considering the sieve of the norm, we encounter the test contained in the categorical imperative as the primary form of prescription following from respect for persons. That is, to incorporate respect for persons in the ethical aim, practitioners ought to consider the categorical imperative when engaging in a certain technical practice. In accordance with Ricoeur (1992, p. 222), it is the second formulation of the categorical imperative that functions as the test: "act in such as way that you always treat humanity, whether in your own person or in the person of any other, never simply as a means, but always at the same time as an end" (Kant 2002, p. 96). Such a test could be integrated in a code of conduct, or in a protocol that would govern the engagement in a certain technical practice. "Following the protocol" would in such a case include concerns expressed by the categorical imperative in order to ensure respect for persons.

Critical solicitude. This refers to the extent to which a practitioner engaging in a technical practice is aware of and responsive to the consideration of vulnerable persons, whose personhood is subject to ambiguity. Vulnerable persons are important to take into account explicitly because they can be affected by conflicts between the norm (e.g., the above-mentioned protocol) and the ethical aim. That is, in some cases the norm might prescribe a certain course of action that conflicts with a practitioner's conception of the good life with and for others. When considering smart border control systems, we argue that such a situation might arise when a traveller is kept at an airport for being stateless (i.e., not having a passport), whose reason for traveling is prosecution by a violent regime. A standard of excellence of the technical practice of securing the border could be that people who are able to make a credible case for being under threat if they are not let through should be let through, even if this means acting against the protocol. A potential problem of smart border control systems is that they do not allow for such a decision based on ambiguity for they simply keep a person from moving when no passport is presented.

The primary form of prescription that follows from these considerations of critical solicitude is the involvement of the *phronimoi* in making a prudent decision regarding ambiguous cases. This means that *expert guidance* is called for in such cases, but expertise should explicitly not be understood as having familiarity in a technique. Rather, expertise should imply a proper level of experience in a practice (e.g., having dealt with refugees trying to cross the border for a long time), and relevant knowledge (e.g., having knowledge of applicable laws and human rights declarations). As argued in Chap. 5, experts should acknowledge that the dilemma they face concerns the principle of respect for persons and that they should aim at finding the mean, which might entail enforcing a dichotomy (e.g., deciding on letting someone pass versus keeping someone at the border). This consideration points to ways in which a proper symbiosis between human and machine might be achieved, for the efficiency gained with installing smart border control systems could be complemented by a greater attention for ways to interrupt the process when ambiguity arises and employing proper experts to deal with ambiguous cases on the spot.

Having considered the good life, with and for others in relation to technical practices, we are able to formulate the eighth set of procedural steps for our method:

- Prescribe forms of community building that promote the relevant virtues of care, friendship and honesty, which incline practitioners to attend to each other reciprocally.
- Prescribe forms of codes or protocols of conduct that ensure practitioners to respect persons while engaging in a technical practice.
- Prescribe forms of expert guidance, which should mediate conflicts arising from ambiguity in dealing with vulnerable persons, arriving at critical solicitude.

In Just Institutions

The ninth and final stage is to the furthest extent concerned with the political dimension, with the realm of institutions. As we argued in Chap. 5, institutions give a temporal dimension to power in common, but also at all times risk being perverted. Therefore, technical practices that cultivate the virtues have to link to the good life *in just institutions*. The institutional dimension cannot be ignored in ethics, because in the form of *domination* it impacts technical practices, be-it regarding the design of technologies involved in those practices, the education provided to promote virtues of practitioners, and so forth. It is only at this stage that the desirability of smart border control systems in general is questioned, for it is at the political level that society can approve of or prohibit particular technologies. We start again with a naïve understanding of just institutions, considered in relation to the virtues of character that shape it, and move through the sieve of the norm of distributive justice to the sense of justice, which demands democratic legitimation.

Equality. This refers to the extent to which practitioners engaging in technical practices relate to unknown others (i.e., the stranger) in accordance with a naïve conception of the good life in just institutions. The relevant *political* virtues of character are, in line with Vallor's heuristic, *justice*, *civility*, and *flexibility*. We can relate these virtues to Ricoeur's theory, first because civility can be considered as the politicised version of the "ethics of argumentation", or the ethics of civil discourse that we encountered before while considering the good life. Justice for Vallor corresponds to equality in Ricoeur's account, for it is concerned with distribution based on the extension of the good towards to common good; flexibility corresponds to justice because it is focuses on arbitration between spheres of justice that are historically situated. We can relate these political virtues to an analysis of technical practice. Concerning smart border controls, we argue that in the related technical practice of crossing the border the virtue of civility, or "making common cause" (Vallor 2016, p. 141) is at stake because the automated system physically separates a person from other people waiting in line, making it difficult to engage in discourse to make common cause when the situation calls for it.

The most prominent form of prescription related to equality in accordance with the virtues of character of civility, justice, and flexibility, is that of *civic education*. Civic education is one of the primary ways of instilling practitioners with political virtues relevant for establishing the for-the-sake-of-which of the technical practices they engage in (cf. Gutmann 1995). This means that practitioners in R&I contexts should receive schooling that explicitly takes into account the political virtues and their relation to particular technical practices, and should have the means for being vigilant in civil life—for instance by being members of trade unions.

The rule of justice. This refers to the extent to which practitioners take principles of distributive justice into account when establishing the for-the-sake-of-which of a technical practice. This places a limit on a technical practice, by ensuring that it does not involve technological configurations that lead to inequalities that are unjustifiable. This aspect of the ethical aim is of particular importance when considering the design of smart border controls. For instance, regarding the practice of "data controlling" of data gathered by smart border control systems, a practitioner might question the abstractions brought about in the process of technological configuration. That is, the system might be triggered to react on certain personalised profiles, thereby affecting the world of action by means of abstract entities. This process would constitute a movement of descending specification, of remote ideals incorporated in an ontology underlying a profiling system (i.e. system bias) towards basic actions (e.g., a traveller having to stay in place). This might violate the rule of justice concerning the technical practice of data controlling, because those affected by the practice could be subject to an arrangement of social and economic inequalities that is not in everyone's advantage (e.g., be halted because of their ethnic profile).

This is the first time we arrive at *legal regulations* as the primary form of prescription in accordance with the rule of justice. This does not imply that only policy changes can follow from considerations of the rule of justice because it can equally well translate into design requirements such as regulations regarding material use, privacy settings in software design, minimum safety requirements, and so forth. However, the focus on regulations does imply that considerations of the rule of justice cannot be merely left to a practitioner's own devices and decisions. Rather, it is the political community that is supposed to formulate and implement regulations that accord with the rule of justice. The role of the practitioner, in this regard, is to (1) pay reasonable respect to the rule of the law that is meant to accord with the rule of justice and to (2) be vigilant in recognising particular moments when the rule of justice is not sufficiently respected and regulations are absent. For this reason, regarding the practitioners, legal regulations should not merely be about the upholding the rule of justice, but also about the empowerment of practitioners in contributing to its enforcement and refinement. As such, it should for instance include regulations for whistle-blowers, protecting their position and encouraging them to voice their concerns.

Sense of justice. This refers to the extent to which practitioners are capable of arbitrating between competing forms of domination, between the governed and the governing and between groups in civil society, in establishing the for-the-sake-of-which of a technical practice. As such, it brings us closest to the significance of the political in the ethics of technology. Again, it does not merely refer to considerations of decision-making in terms of constitutional or legal procedures, but also relates to design choices, such as the channeling of political speech on Internet forums, the possibility of users to flag digital content and co-decide on what is done with it, and so forth. Design is an increasingly pressing issue in particular because digital technologies are increasingly developed for political purposes: for voting, organising public debates, contracting and decision-making. Considering smart border controls, we argue that arbitration is needed with regard to the domination of states over the people travelling through airports. For instance, smart border controls could arbitrarily halt people from a particular nationality, thereby arbitrating between the powers of the governed (the travellers) vis-à-vis the power of the governing (the state). As such, smart border controls would compromise the ability of civilians to challenge the domination embedded in the technical practice of securing the border, as for instance happened during protests in the United States at airports, where border police officers in some cases refused to carry out the federal executive order signed by the president that ordered them to halt people from certain nationalities (Gambino and Siddiqui 2017). This challenge can in some cases lead to a legitimate plea and decision for banning a particular technology altogether.

The primary form of prescription following from considerations of the rule of justice is captured under the notion of *democratic decision-making*. Democratic decision-making is a complex and contested issue that perhaps will never be fully resolved (we will not here engage in an elaboration of different democratic decision-making models in democratic theory), but practitioners are able to at least take it into account in technical practices to the furthest extent possible. That is, by tentatively agreeing on ways in which democratic decision-making can be organised, relevant stakeholders could implement agreed-upon democratic principles in formal policies and procedures as well as in requirements for technology design. Such principles could for instance guide decisions with regard to user-agency in creating digital contents on public forums on the Internet, provide the governed (i.e. users) with the means to challenge the governing (i.e. Internet companies), and put conflicts forward to public debate (i.e. to the relevant legislators). Such a public debate should be organised in a two-fold way: both (1) by questioning the particular technical practice, informed either by norms of good governance or novel or refined norms, and by (2) questioning the very process shaping the decision-making.

Having considered the good life in just institutions related to technical practices, we have come to the point that we can formulate a ninth set of procedural steps:

- Prescribe forms of civic education that promote the relevant political virtues of justice, civility and flexibility, which incline practitioners to attend to the establishment of equality.
- Prescribe forms of legal regulations that ensure practitioners to respect the rule of justice and enhance their ability to challenge violations of this rule.
- Prescribe forms of democratic decision-making that mediate conflicts by moving them forwards to public debate, according to received and novel norms of good governance while at all times being critical of and questioning the process of decision-making itself, thereby establishing a sense of justice.

Conclusion

The above-mentioned nine steps together form the proposed novel method for practising ethics in R&I. What they make explicit is that practising ethics is not limited to evaluating the "pros" and "cons" of a particular technology, or advising on the desirability of a technology. These aspects are important, but far from sufficient for striving for technical practices that cultivate the virtues for the sake of the good life, with and for others, in just institutions. To cultivate such technical practices the well-being, sense of community, and civic education of practitioners needs to be taken into account. For instance, an engineer without any sense of justice cannot be expected to engage in the technical practice of "designing a smart border control system" in a virtuous manner, which might in some cases imply refraining from the practice altogether. Furthermore, and to some extent as a counter-balance to Aristotelian virtue ethics, technical practices that cultivate the virtues cannot persist without personal norms (e.g., through an ethical oath), communal norms (e.g., through codes of conduct) and political norms (e.g., through regulations), imposing limits on those practices. Finally, the inevitability of conflicts between norms and a refined understanding of the good life calls for the necessity of structured dialogue, expert guidance and democratic decision-making as prudent means to establish what one might call "best practices".

The critical reader might comment that the notion of "design" in the method is somewhat misleading, because no specific guidance for technology design is provided. In response to this objection, we reply that by changing the focus of the method from "technology" to "technical practice", the focus and meaning of "design" is also changed by encompassing a technical practice as a whole. That is, "ethical design" of for instance a smart border control system does not *merely* refer to changing the specifications of the actual thing, for instance its sensors, its camera, the software it runs, and so forth. More importantly, it refers to the design of the related technical practices of for instance "crossing the border" or "securing the border". As such, it never suffices to merely change the design of the particular technology to make it "ethical", but it always also requires attending to the standards of excellence of the practices in question, the norms that shape these and the notion of the good life that practitioners can relate to in doing their day-to-day work.

One could moreover object that the procedural steps are still very general and do not yet sufficiently answer the exact question of "how" they can be implemented in practice. For instance, the procedural step concerned with decision-making does not indicate what *kind* of decision-making should be used in technical practices (i.e. representative democracy, direct democracy, citizen councils, and so forth). We provide two answers as a response. The first answer concerns the role of the philosopher in formulating a method like the one proposed here. We argue that in order for procedural steps to be responsive to a particular technical practice, they require the input from those engaging in and knowledgeable of that practice. This implies that in developing procedural steps further, to make them responsive to for instance the technical practice of designing technologies, a philosopher would need to engage with practitioners, with legislators that have experience in drafting related legislation, and so forth. In other words, further development and refinement of the procedural steps would require an inter- and multidisciplinary effort and cannot solely be a philosophical effort.

The second answer concerns the character of more specific procedural steps, which we argue would take us from an understanding of method to an understanding of practical "tools", such as the Ethics Canvas (Reijers et al. 2018). A philosophical method, we argue, should present a general way of doing things that is backed up by a philosophical theory, but should not present definite answers to the "how" question in the form of a tool, understood as a "thing" that enables the in-order-to relation (e.g., "this tool can be used in-order-to formulate a code of ethics"). This is not to say that such tools are unimportant. On the contrary, they are a necessary component of practising ethics in R&I, *but* at the same time should always be subordinate to the normative guidance that is linked to existing standards of excellence in the community, the ethical aim(s), and the philosophical theory and method from which they originate or which justifies them. In other words, tools for practising ethics should not become ends in themselves and should always be challenged on the basis of ends that are present in, and derived from, communal standards, ethical aims, and philosophical critique and convictions.

References

Adamic, L. A., & Glance, N. (2005). The Political Blogosphere and the 2004 U.S. Election. *Proceedings of the 3rd International Workshop on Link Discovery—LinkKDD '05*, 36–43. https://doi.org/10.1145/1134271.1134277.

Arendt, H. (1964). *Eichmann in Jerusalem: A Report on the Banality of Evil.* New York: Viking Press. https://doi.org/10.1525/california/9780520220560.001.0001.

Aristotle. (1999). In T. Irwin (Ed.), *Nicomachean Ethics.* Indianapolis, IN: Hackett Publishing Company.

Asif, M., & Krogstie, J. (2012). Taxonomy of Personalization in Mobile Services. *Proceedings of the 10th IADIS International*…, (January), 343–350. Retrieved from http://www.researchgate.net/publication/258847751_Taxonomy_of_Personalization_in_Mobile_Services/file/60b7d5293d9f45c89a.pdf.

Blok, V. (2013). The Power of Speech Acts: Reflections on a Performative Concept of Ethical Oaths in Economics and Business. *Review of Social Economy, 71*(2), 187–208. https://doi.org/10.1080/00346764.2013.799965.

Bloor, M., Frankland, J., Thomas, M., & Robson, K. (2001). *Focus Groups in Social Research.* London: SAGE Publications. https://doi.org/10.4135/9781849209175.

Bohn, J., Coroama, V., Langheinrich, M., & Mattern, M. (2005). Social, Economic, and Ethical Implications of Ambient Intelligence and Ubiquitous Computing. *Ambient Intelligence, 10*(5), 5–29. https://doi.org/10.1007/3-540-27139-2_2.

Bouzeghoub, M., & Kostadinov, D. (2007). *Data Personalization: A Taxonomy of User Profiles Knowledge and a Profile Management Tool. Rapports de recherche du laboratoire PRiSM,[en ligne].*

Brey, P. (2000). Disclosive Computer Ethics: The Exposure and Evaluation of Embedded Normativity in Computer Technology. *Computers and Society, 30*(4), 10–16.

Clavell, G. G. (2017). Protect Rights at Automated Borders. *Nature, 543*(7643), 34–36. https://doi.org/10.1038/543034a.

De Vries, K. (2010). Identity, Profiling Algorithms and a World of Ambient Intelligence. *Ethics and Information Technology, 12*(1), 71–85. https://doi.org/10.1007/s10676-009-9215-9.

Ezra, M. (2007). The Eichmann Polemics: Hannah Arendt and Her Critics. *Democratiya, 9*(Summer), 141–165.

Friedman, B., & Kahn, P. (2002). Value Sensitive Design: Theory and Methods. *University of Washington Technical,* (December), 1–8. https://doi.org/10.1016/j.neuropharm.2007.08.009

Gadamer, H.-G. (2013). In J. Weinsheimer & D. G. Marshall (Eds.), *Truth and Method.* London: Bloomsbury.

Gambino, L., & Siddiqui, S. (2017). Thousands Protest Against Trump Travel Ban in Cities and Airports Nationwide. Retrieved May 13, 2018, from https://www.theguardian.com/us-news/2017/jan/29/protest-trump-travel-ban-muslims-airports.

Gutmann, A. (1995). Civic Education and Social Diversity. *Ethics, 105*(3), 557–579.

Hargittai, E. (2007). Whose Space? Differences Among Users and Non-users of Social Network Sites. *Journal of Computer-Mediated Communication, 13*(1), 276–297. https://doi.org/10.1111/j.1083-6101.2007.00396.x.

Heidegger, M. (1977). *The Question Concerning Technology*. New York, NY: Garland Publishing.

Ihde, D. (1979). *Technics and Praxis* (Vol. 33). Dordrecht, Netherlands: D. Reidel Publishing Company.

Jovchelovitch, S., & Bauer, M. W. (2000). Narrative Interviewing. In M. W. Bauer & G. Gaskell (Eds.), *Qualitative Researching with Text, Image and Sound: A Practical Handbook*. London: SAGE Publications.

Juels, A., Molnar, A., & Wagner, D. (2005). Security and Privacy Issues in ePassports. *Security and Privacy for Emerging Areas in Communications Networks, 2005*, 74–88. https://doi.org/10.1109/SECURECOMM.2005.59

Kant, I. (2002). In J. B. Schneewind, M. Baron, S. Kagan, & A. W. Wood (Eds.), *Groundwork for the Metaphysics of Morals* (Vol. 47). New Haven, CT: Yale University Press.

Kaplan, D. M. (2003). *Ricoeur's Critical Theory. SUNY Series in the Philosophy of the Social Sciences*. New York, NY: SUNY Press. https://doi.org/10.1017/CBO9781107415324.004.

Latour, B. (2013). In C. Porter (Ed.), *An Inquiry into the Modes of Existence: An Anthropology of the Moderns*. Cambridge, MA: Harvard University Press.

Lichterman, P. (1995). Piecing Together Multicultural Community: Cultural Differences in Community Building among Grass-Roots Environmentalists. *Social Problems, 42*(4), 513–534. https://doi.org/10.1525/sp.1995.42.4.03x0130z.

Liebowitz, S. J., & de Meuse, K. P. (1982). The Application of Team Building. *Human Relations, 35*(1), 1–18. https://doi.org/10.1177/001872678203500102.

Lycett, M. (2013). "Datafication": Making Sense of (Big) Data in a Complex World. *European Journal of Information Systems, 22*(4), 381–386. Retrieved from http://v-scheiner.brunel.ac.uk/handle/2438/8110.

Miller, J. I. (2005). Don't Be Evil: Gmail's Relevant Text Advertisements Violate Google's Own Motto and Your E-Mail Privacy Rights. *Hofstra Law Review, 33*, 1.

Reijers, W., Koidl, K., Lewis, D., Harshvardhan, J. P., & Gordijn, B. (2018). Discussing Ethical Impacts in Research and Innovation: The Ethics Canvas. In D. Kreps, G. Fletcher, & M. Griffiths (Eds.), *IFIP Advances in Information and Communication Technology, HCC13, Poznan, Poland, September 19-21, 2018*. Cham: Springer International Publishing.

Reijers, W., Wright, D., Brey, P., Weber, K., Rodrigues, R., O'Sullivan, D., & Gordijn, B. (2017). Methods for Practising Ethics in Research and Innovation:

A Literature Review, Critical Analysis and Recommendations. *Science and Engineering Ethics, 24*, 1437–1481. https://doi.org/10.1007/s11948-017-9961-8.

Ricoeur, P. (1992). In K. Blamey (Ed.), *Oneself as Another*. Chicago, IL: University of Chicago Press.

Roosendaal, A. (2014). Digital Personae and Profiles as Representations of Individuals. In M. Bezzi, P. Duquenoy, S. Fischer-Hubner, M. Hansen, & G. Zhang (Eds.), *Privacy and Identity Management for Life* (pp. 226–236). Heidelberg: Springer.

Schubert, C. (2015). *Christian Schubert on the Ethics of Public Nudging: Autonomy and Agency On the Ethics of Public Nudging: Autonomy and Agency* (No. Joint Discussion Paper Series in Economics No. 33-2015). University of Kassel.

Vallor, S. (2016). *Technology and the Virtues: A Philosophical Guide to a Future Worth Wanting*. Oxford: Oxford University Press.

Weigel, M. (2017). Coders of the World, Unite: Can Silicon Valley Workers Curb the Power of Big Tech? Retrieved May 13, 2018, from https://www.theguardian.com/news/2017/oct/31/coders-of-the-world-unite-can-silicon-valley-workers-curb-the-power-of-big-tech.

7

Conclusion: Hermeneutic Ethics for the Digital Age

In this concluding chapter, we do not merely summarise the findings of *Narrative and Technology Ethics*, but also reflect on how it might open up new directions of philosophical research; including investigations on the limits of narrative for the understanding of technical practice. In the introduction, we formulated the thesis that technologies, similar to texts, novels, and movies, "tell stories" by configuring characters and events in meaningful syntheses. We did not engage with the thematic directly, but instead started our investigation with the notion of technical practice; the focal point of technology ethics. We saw that technical practice opens up the problem of the "nesting of finalities", and that this problem is addressed through narrative. That is, technical practices involve a nesting of basic actions with regard to life plans and the unity of life. Technologies play a role as co-authors in this nesting of finalities, by mediating the movement of "ascending complexification" and "descending specification", which links basic actions to more or less distant ideals (and therefore, with institutions). This mediation can be characterised according to the basic aspects of textuality, literacy, temporality, and distancing. Finally, we faced the distinction between a "mere" technical practice, or perhaps rather the training of a technique, and a "virtuous" technical

W. Reijers, M. Coeckelbergh, *Narrative and Technology Ethics*,
https://doi.org/10.1007/978-3-030-60272-7_7

practice, which demands a practitioner to be serious in his or her handling of technologies. Genuine virtuous technical practice, moreover, has to be aligned with the ethical aim; meaning that they aim at the good life, with and for others.

In drawing this rather theoretical understanding of technical practice back to the concrete "ground" of research and innovation (R&I) practices, we translated our account into a method for practising ethics. This method provides a comprehensive "map" of things that practitioners need to take into account if they want to engage in *responsible* R&I. This traverses all the way from cultivating the virtues of the individual, through interventions like mentorship, improving relations with others, via interventions such as expert guidance, up to bettering institutional design, through interventions like changes in democratic decision. Even though this method stays away from proposing practical tools for integrating ethics in R&I, it has informed such tools. A prominent example is the "Ethics Canvas" (Reijers et al. 2018). This canvas is a visual-narrative tool that R&I practitioners can use, both as a physical canvas and as an online platform, to discuss the potential ethical impacts of their work, and thereby reveal relevant narrative structures.

Narrative and Technology Ethics thus offers a new, hermeneutic approach to technology ethics; that aims to be truly comprehensive. Yet, further explorations of for instance the political dimension of R&I will be required, as well as an elaborate investigation of the limitations of a narrative approach, in order to get to a place from where we can face the major challenges of our time. As such, the method developed in this book is merely a small step in contributing to a world in which the impact of R&I in the context of Covid19, cyberwar, and the looming danger of climate change can be subjected to a "responsible way" of engaging in technical practices. In what follows, we will briefly summarise the main contributions of the book and look ahead towards future directions of research.

Summary

Narrative & Technology Ethics draws on, and responds to, three bodies of theory. The first one is *Ricoeur's narrative theory*, which draws from Aristotle and influenced by Arendt and MacIntyre. This enabled us to develop a narrative theory of technology, which, in line with philosophy of technology's emphasis on the more-than-instrumental character of technology, shows how technology is not merely the object of our narratives but also configures and co-authors our actions and narratives, including the narratives of our lives. A new, hermeneutic account of technical practice was proposed, in a three-stage way: starting (1) with the delineation of the concept of technical practice according to the movement from basic action, through life plans, to the narrative unity of life; which is linked to (2) a theory of technological mediation that explicates how technologies mediate practices according to the model of emplotment, and (3) completed through a notion of the ethical aim, which provided an understanding of virtuous technical practice. We argued that, and showed how, technologies mediate and configure our actions and experience, including our understanding of temporality and distancing.

Going beyond Ricoeur, our effort delivered a hermeneutics that is not for the most part concerned with text (Ricoeur) or theatre (Aristotle), but extends to technology. Next to contributing to Ricoeur-based scholarship by interpreting him as a philosopher of mediation (Chap. 3) and by developing his narrative theory for use with regard to technology, this theory of technology also helps to give more substance to the project of an hermeneutics of technology, including hermeneutics of digital technologies (cf., Romele 2019). We now have a framework that enables us to analyse the narrative roles of technology. This was missing in Ihde's posthermeneutics and in mediation theory. In addition, a narrative theory of technology inspired by Ricoeur does not only pay attention to narrative as such, but also helps us to fill the gaps in philosophy of technology after the empirical turn identified in Chap. 2: it delivers a theory that is more sensitive to the social, public dimension of technology's mediations (the social), incorporates concerns for linguistic mediations, and reveals the temporal dimension of human and technological agency.

This led us to the second body of knowledge: *philosophy of technology*. Next to Postphenomenology and mediation theory, we also responded to Vallor's virtue ethics of technology. Her work successfully showed how thinking about the virtues can help with the ethical evaluation of technologies. However, Vallor also insufficiently acknowledges the non-instrumental, active, and mediating role of technology, while she discusses other-regarding virtues more needed to be said about community and about what we later called 'with and for others', and her work did not sufficiently account for the way virtues are linked to a practice. The latter is important since we wanted a narrative ethics of technology that is not only about technologies (and narratives of course) as such but also about technological practice.

This is why, thirdly, we also turned to *virtue ethics*, and in particular MacIntyre's version of virtue ethics, again based on Aristotle. MacIntyre's ethics suited our purpose since it offers a theory of practice and allows us to make a direct link from the virtues to practice. MacIntyre's notion of practice is linked to the concepts of (a) narrative and (b) community (next to moral norms), since the meaningful exercise of the virtues is and must be embedded in these wider wholes. While there are some limitations to MacIntyre's theory, our critical use of it delivered an ethics of technology that allows for the role of narrative and is less individualistic and more firmly rooted in the notion of a practice and the relevant communities than Vallor's. Moreover, we are also able to distinguish (in Chap. 5) between practices that cultivate the virtues and those that do not. Next to the addition of moral norms, this counters the relativism objection to which MacIntyre's account is all too vulnerable. Inspired by Ricoeur's interpretation of Aristotle, in particular the notion of the ethical aim and his "little ethics" that tries to reconcile the deontological tradition (morality) with the Aristotelian, teleological one (phronesis) and Hegel's *Sittlichkeit*, we put technical practice in the context of the question concerning the good life, with and for others, and with regard to just institutions. The cultivation of virtue in and through technical practices is thus linked not only to ways of doing and narratives within a particular community, but also to moral and political principles. In this way, we bring back discussion about the ends of technology to philosophy and ethics of technology, without however regressing into an instrumental understanding of technology itself.

Future Directions

What does this mean for technology ethics and responsible innovation? In Chap. 6 we translated our narrative ethics of technology, understood as an ethics of technical practice, into a concrete method for practising ethics in R&I. We illustrated the application of this method in our discussion of automated border control systems, describing narrative structures and movements of technological emplotment. We also analysed technical practices according to the way they link to standards of excellence, life plans, and the narrative unity of life, and according to their relation to the good life and justice. Thus, we did not only deliver a descriptive-hermeneutic narrative theory of technology (responding to Ricoeur and MacIntyre) and a further development of virtue ethics of technology (Vallor), but also a novel and original method for evaluating technologies in R&I contexts. We thus developed a framework that meets two aims, a hermeneutic one and a normative, prescriptive one. In the remainder of this concluding chapter, we outline further potential directions for research and implementation of the proposed framework.

Aristotle and the Limit of Narrativity

First, since the Aristotelian view continues to inform philosophy of technology, including the present work, it is worth to turn, or rather re-turn to Aristotle, directly or through the work of Heidegger, Ricoeur, and other thinkers in hermeneutics. In this book we could only cursorily engage with Aristotle. Yet, he is a background figure in the works of our main intellectual sources (Ricoeur and MacIntyre). Renewed, direct engagement with Aristotle could deliver new, original insights into hermeneutics, including hermeneutics of technology. For example, while contemporary philosophy of technology should reject his naïve dualism with regard to human-technology relations, his *Poetics*—read in conjunction with his work on ethics and politics—may well be a fertile ground for further reflection on the meaning of technology.

If a more performative approach to hermeneutics of technology would be welcome, then a discussion of Aristotle's view of theatre and narrativity (and its broader reception history) might be rendered relevant to our project in ways not yet fully covered by Ricoeur. For example, in Book XV of the *Poetics* Aristotle does not allow the famous theatrical technology "deus ex machina" to participate in the plot: the gods should not intervene, they can only cast judgement. Aristotle did not want improbable events. In general, Aristotle does not see a big role for technology in the narrative of the theatre (tragedy), nor does Ricoeur. The plot is all about human characters and events. In our book we have questioned this anthropocentrism. Moreover, technology might well be a way to bring back the improbable and unexpected. If technology mediates and configures human narratives, as we argued in this book, this does not necessarily imply that these narratives become "mechanical". Like all narratives that involve humans there is plenty of room for the unexpected and sometimes improbable to happen—with or without gods.

At the same time, we should also question the limits of the Aristotelian, narrative approach. In this book, we took a rather "positive" outlook; assuming semantic optimism in our approach of technical practice and the phenomenon of technological mediation. That is, we paid in particular attention to how technical practices are intelligible to human being, but less so to the unintelligibility of technical practices. On the flipside, following the Aristotelian tradition and its emphasis on the intelligibility of mimesis has its limits. For instance, Romele (2019) discusses how digital machines develop a certain autonomy, gaining a form of productive imagination that has an effect in the world but nonetheless is for a large part inaccessible to human interpretation. Similarly, scholars like Dupont (2017) stress how technologies like cryptography obscure or even destroy meaning, which throws doubt on a "mimetic" viewpoint. Further developments of the hermeneutic approach should take into account the anti-hermeneutical or anti-interpretative qualities of technologies; which therefore illuminate the limits of narrative, and perhaps even the limits of the notion of technical practice.

Developing a Hermeneutics of Technology

Second, we propose further develop a hermeneutics of technology (or: hermeneutic philosophy of technology) as a relatively new field, using narrative but also other approaches. It could be interesting to connect to other recent work on hermeneutics of technology, for example by Romele (2019), and Coeckelbergh's (2017, 2019) approach to understanding the meaning of technology in terms of use and performance. For the latter, use is interpreted in line with Wittgenstein (1958) as involving (and contributing to) what he calls "technology games" and forms of life. The meaning of particular technologies is not given in the object (the artefact) but is constituted through use with activities and the related social and cultural frameworks available. This suggests a holistic view of the hermeneutics of technology. And the concept of performance—used by Coeckelbergh as a metaphor—enables to bring in the aspects of embodiment, but also the social and temporality.

Engagement with the idea of technology as performance raises the question of how Coeckelbergh's understanding of the temporal aspect of technological performances compares with the understanding of temporality of technological mediation we proposed. Also, how does the social-cultural dimension of technology in terms of Wittgensteinian games and forms of life compare to the community dimension of the notion of technical practice we developed by using MacIntyre? Whereas MacIntyre tends to put agency in the context of narrativity at the level of life plans and particular communities, we have argued for a broader application of narrativity that extends to wider society (including questions of distributive justice in that wider society), one which seems compatible with the notion of forms of life. Moreover, we could explore how our work compares to Romele's approach to the digital understood (in a more sociological way perhaps) as a technological milieu and a cultural phenomenon, including not only visible effects but also invisible blending with the life-world? Are there ways to integrate these different perspectives on the hermeneutics of technology?

For example, one could elaborate the notion of a *performative technical practice*. The idea is that technical practice is not only hermeneutically embedded in narratives and forms of life, but is also performative in the sense of being embodied and involving bodily movement, being related to the social in various ways, and being temporal also in a concrete, kinetic sense. More work needs to be done on the role of the body and movement in a hermeneutics of technology. Furthermore, Aristotle needs not be the only source for thinking about praxis. One could also further engage with Marx's or Bourdieu's notion of praxis and explore what this could mean for narrative hermeneutics of technology (with regard to Bourdieu, Romele's work is again relevant). As we have seen, Bourdieu's notion of habitus was crafted to link the individual to its social context; conceptualising this link remains important with regard to the project of hermeneutics of technology, understood as a hermeneutics of practice. How does the notion of habitus connect to the notions of practice and narrative?

A further suggestion is to connect the proposed narrative approach more with ongoing work in science and technologies studies (STS), in the vein of Latour. In the Latourian direction, one could bring our Ricoeurian framework into dialogue with contributions to narrative theory by Greimas, Latour, and Czarniawasma. With regard to this dialogue, it would be also interesting to unpack the theatre metaphor that underlies Greimas's approach and—via Aristotle—Ricoeur's hermeneutics. In the end, understanding and evaluating technology may have more to do with understanding and evaluating theatre than is commonly presupposed: from the ancient theatre with its machinations to the AI performances of today.

The Politics of Responsible Innovation

Third, future research could put more emphasis on the political aspects of a hermeneutic approach to responsible innovation, which could for instance be developed by drawing on Dewey and democratic theory. How can a democratic deliberation on the ethical impacts of technology take place?

Ethics of technology is the first, obvious field of application. After developing our approach, it is now a task to use it for interpreting and evaluating all kinds of new and emerging technologies, including for instance artificial intelligence and robotics. For example, how does ethics of AI fare when it no longer considers AI as a technological artefact, code, or "thing", but as a technological practice? And what kind of hermeneutic mediations is AI involved in? What does it do in terms of emplotment, how does it shape the narrative structure of our lives and our societies? How does AI organise and re-organise characters and events? Links could be made not only to responsible innovation frameworks, as we proposed in the previous chapter, but also to work on AI ethics and methods of technology assessment (e.g., Coeckelbergh 2020).

But we propose to also consider political philosophy and related fields, and to connect to other strands in the history of philosophy. For example, our approach could benefit from dialogue with other intellectual traditions. Consider for instance the work of Dewey. Dewey's pragmatism influenced Ihde's postphenomenology, with its use-oriented direction. But it could be also used to think about democratic participation in the context of responsible R&I, and to further elaborate the social aspects of technological mediations understood in terms of technical practices. The meaning of technology is then not only related to personal lives or communities, but also to efforts at problem solving and experimentation at societal level. More generally, the method we presented could be further developed by relating it not only to responsible innovation, as we did in the previous chapter, but also to questions in political philosophy.

Next to questions regarding democracy, we could also consider the relation between virtue and governance. In contrast to much modern political-philosophical thinking, there may well be a constitutive relation between virtue and governance. The discourse on responsible innovation need not be restricted to talk about "governance" in an abstract way; we can also ask the questions about individual agency and virtue with regard to responsible innovation. For instance, Blok et al. (2016) have asked what the role of individual virtuous competence is in dealing with sustainability problems. While the previous chapter has said much about a hermeneutic approach to responsible innovation, more needs to be said about the role of virtue, in relation to structures of governance. More generally, how are virtue and politics related in responsible innovation and, broader, in a hermeneutic approach to responsible innovation and technology ethics?

Conclusion

Narrative & Technology Ethics offers a narrative theory of technology and a narrative virtue ethics of technology. It brings a unique perspective to contemporary philosophy of technology, which as an empirically oriented project that takes distance from the Heideggerian tradition, from hermeneutics, and sometimes from the philosophical tradition as a whole, is all too often insensitive to the linguistic, social, and temporal dimensions of technological practice—and indeed does not even use the notion of practice. The book has shown that there is much to be gained in this respect, while acknowledging that more work is needed to further elaborate this narrative approach to technology and to technology ethics, and more generally hermeneutic philosophy of technology. We hope that more researchers will contribute to the fascinating goal of developing this area, not only for the sake of giving this specific field a place in academia as such, but also and especially for the sake of making sure that we have the right conceptual tools to adequately understand and evaluate (our relation to) new technologies understood as practices and as entangled with narrative in various ways.

References

Blok, V., Gremmen, B., & Wesselink, R. (2016). Dealing with the Wicked Problem of Sustainability: The Role of Individual Virtuous Competence. *Business and Professional Ethics Journal, 34*(3): 297–327.

Coeckelbergh, M. (2017). *Using Words and Things: Language and Philosophy of Technology*. New York: Routledge.

Coeckelbergh, M. (2019). *Moved by Machines: Performance Metaphors and Philosophy of Technology*. New York; Abingdon; Oxon: Routledge.

Coeckelbergh, M. (2020). *AI Ethics*. Cambridge, MA: The MIT Press.

Dupont, Q. (2017). Blockchain Identities: Notational Technologies for Control and Management of Abstracted Entities. *Metaphilosophy, 48*(5), 634–653. https://doi.org/10.1111/meta.12267.

Reijers, W., Koidl, K., Lewis, D., Harshvardhan, J. P., & Gordijn, B. (2018). Discussing Ethical Impacts in Research and Innovation: The Ethics Canvas. In D. Kreps, G. Fletcher, & M. Griffiths (Eds.), *IFIP Advances in Information and Communication Technology, HCC13, Poznan, Poland, September 19–21, 2018*. Cham: Springer International Publishing.

Romele, A. (2019). *Digital Hermeneutics*. London: Routledge. https://doi.org/10.4324/9780429331893.

Wittgenstein, L. (1958). *Philosophical Investigations*. Oxford: Basil Blackwell Ltd.

Index[1]

A

[1] Note: Page numbers followed by 'n' refer to notes.

W. Reijers, M. Coeckelbergh, *Narrative and Technology Ethics*,
https://doi.org/10.1007/978-3-030-60272-7

B

C

R

S

The manufacturer's authorised representative in the EU is Springer Nature Customer Service Centre GmbH, Europaplatz 3, 69115 Heidelberg, Germany. If you have any concerns regarding our products, please contact ProductSafety@springernature.com

Printed and bound by CPI Group (UK) Ltd, Croydon, CR0 4YY

30/09/2026

02225494-0002